Praise for *The Field*

"This is an important book and should be widely read. . . . It stretches the imagination, making a good case that we are on the verge of another revolution in our understanding of the universe—perhaps even greater than the one that heralded the Atomic Age."—Arthur C. Clarke

"Written with the clarity, grace, and elegance that are her trademark, Lynne McTaggart's *The Field* is a preview of third-millennium science and how it will touch the lives of every person on earth. This book liberates consciousness and restores it to its majestic and rightful position as a causal power in the universe. McTaggart's book should come with a warning: MAY FOREVER CHANGE YOUR WORLDVIEW."

—Larry Dossey, M.D., author of *Healing Words, Reinventing Medicine,* and *Healing Beyond the Body*

"A fascinating and excellent presentation about the true nature of life that we need to be aware of and accept." —Bernie Siegel, M.D., author of *Love, Medicine & Miracles* and *Prescriptions for Living*

"One of the most powerful and enlightening books I have ever read."

—Wayne W. Dyer

"The Force is with us, even if mainstream science is taking its time to get up to speed in acknowledging it. Fortunately, as UK-based scientific-medical writer Lynne McTaggart highlights in *The Field,* frontier scientists have been performing experiments that prove the reality of a realm that hitherto had been the domain of Eastern mystics and Western witches who met their demise at the stake. . . . The vast scope of this book lifts the veil on a state of being that is our birthright." —*Nexus*

"Fascinating, provocative, and highly readable. . . . One of the most thought-provoking reads of [the year.]" —*The Ecologist*

"[This book] will help us to understand the visual human aura, the human memory, the power to heal, the human spirit, and many other fascinating aspects of this thing we call 'human.' Lynne McTaggart has done us a great favor by her extensive reporting on this wondrous topic." —*Journal of Energy Medicine*

Robert Taylor

About the Author

LYNNE McTAGGART is an American investigative journalist whose books include *The Baby Brokers: The Marketing of White Babies in America* and *Kathleen Kennedy: Her Life and Times*. In the mid-eighties she moved to England and founded a newsletter, *What Doctors Don't Tell You,* which today has a six-figure circulation. *What Doctors Don't Tell You* was published as a book by Avon in 1999.

THE FIELD

THE QUEST FOR THE SECRET FORCE OF THE UNIVERSE

Lynne McTaggart

An Imprint of HarperCollins*Publishers*

The author is grateful to Robert G. Jahn and Brenda J. Dunne for permission to quote from their book *Margins of Reality* (New York: Harcourt Brace Jovanovich, 1987).

First published in Great Britain in 2001 by HarperCollins Publishers.

The first U.S. edition of this book was published in 2002 by HarperCollins Publishers.

First Quill edition published 2003.

The Library of Congress has catalogued the hardcover edition as follows:

McTaggart, Lynne.
The field : the quest for the secret force of the universe / Lynne McTaggart.— 1st ed.
p. cm.
Includes bibliographical references and index.
ISBN 0-06-019300-X
1. Alternative medicine. 2. Mind and body. I. Title.

R733 .M396 2002
615.5—dc21 2002017348

ISBN 0-06-093117-5 (pbk.)

03 04 05 06 07 ❖/RRD 10 9 8 7 6 5 4 3 2 1

FOR CAITLIN
YOU NEVER WERE ALONE

Physics may be about to face a revolution similar to that which occurred just a century ago . . .

Arthur C Clarke, 'When Will the Real Space Age Begin?'

If an angel were to tell us about his philosophy . . . many of his statements might well sound like 2 x 2 = 13.

Georg Christophe Lichtenberg, *Aphorisms*

Acknowledgments

THIS BOOK GOT STARTED eight years ago, when in the course of my work I kept bumping up against miracles. Not miracles in the ordinary sense of the term, where the seas part or loaves of bread exponentially multiply, but miracles, nonetheless, in their utter violation of the way we think the world works. The miracles that I came across had to do with hard scientific evidence concerning methods of healing that flout every notion we have about our own biology.

I discovered, for instance, some good studies about homeopathy. Randomized, double-blind, placebo-controlled studies – the gold standard of modern scientific medicine – showed that you could take a substance, dilute it so much that not a single molecule of the substance remained, give this dilution – now nothing more than water – to a patient and the patient would get better.[1] I discovered similar studies for acupuncture; poking the skin with fine needles at certain points of the body along so-called energy meridians was shown, in good solid studies, to work for certain conditions.

As for spiritual healing, although some studies were of poor quality, a number were good enough to indicate that something interesting was going on here, and there might be something more to distant healing than just a placebo or feel-good effect. In many of the studies, patients didn't even know anybody was attempting to heal them. Nonetheless, here was evidence that certain people could concentrate on a patient at a distance, and somehow that person would get better.

These discoveries left me with wonder but also profoundly unsettled. All these practices were based on an entirely different paradigm of the human body from that of modern science. These were medical systems which purported to work on 'energetic levels', but I kept wondering precisely which energy it might be that they were talking about.

In the alternative community, words like 'subtle energy' were often bandied about, but the debunker in me was left dissatisfied. Where was this energy coming from? Where did it reside? What was so subtle about it? Were there such things as human energy fields? And did they account not only for these alternative forms of healing but also for many of life's mysteries that couldn't be explained? Was there an energy source that we didn't really understand?

If something like homeopathy worked, it upended everything we believe about our physical and biological reality. One of the two – homeopathy or standard medical science – had to be wrong. Nothing less than a new biology, a new physics, seemed necessary to embrace what appeared to be true about so-called energy medicine.

I began a personal quest to find out whether any scientists were doing work that suggested an alternative view of the world. I traveled to many areas around the globe, meeting with physicists and other top frontier scientists in Russia, Germany, France, England, South America, Central America and the US. I corresponded with and phoned many other scientists in other countries. I attended conferences at which radically new findings were presented. In the main, I decided to stick to scientists with solid credentials operating according to rigorous scientific criteria. Enough speculation had already been made in the alternative community about energy and healing, and I wanted any new theories to be firmly rooted in what was provable, mathematically or experimentally – precise equations, a real physics to grapple with and understand. As I'd looked to science to prove conventional or alternative medicine, so I wanted the scientific community to provide me with – in a sense – a new science.

Once I began digging, I discovered a small but cohesive community of top-grade scientists with impressive credentials, all doing some small aspect of the same thing. Their discoveries were incredible. What they were working on seemed to overthrow the current laws of biochemistry and physics. Their work not only offered an explanation of why homeopathy and spiritual healing might work. Their theories and experiments also compounded into a new science, a new view of the world.

The Field has largely resulted from interviews with all the major scientists mentioned in the book, plus a reading of their major published work.

These include chiefly: Jacques Benveniste, William Braud, Brenda Dunne, Bernhard Haisch, Basil Hiley, Robert Jahn, Ed May, Peter Marcer, Edgar Mitchell, Roger Nelson, Fritz-Albert Popp, Karl Pribram, Hal Puthoff, Dean Radin, Alfonso Rueda, Walter Schempp, Marilyn Schlitz, Helmut Schmidt, Elisabeth Targ, Russell Targ, Charles Tart and Mae Wan-Ho. I received a herculean amount of help and support from each one of them in person, by telephone and through the post. Most of the individual scientists were involved in multiple interviews – many ten interviews or more. I am indebted to them for consenting to so many consultations and for allowing me to check facts laboriously. They put up with my constant intrusion and also my ignorance, and their assistance has been incalculable.

I must especially thank Dean Radin for educating me in statistics, Hal Puthoff, Fritz Popp and Peter Marcer for what amounted to a course in physics, Karl Pribram for an education in brain neurodynamics and Edgar Mitchell for sharing the most up-to-date developments.

I am also grateful to the following, all of whom I spoke or corresponded with: Andrei Apostol, Hanz Betz, Dick Bierman, Marco Bischof, Christen Blom-Dahl, Richard Broughton, Toni Bunnell, William Corliss, Deborah Delanoy, Suitbert Ertel, George Farr, Peter Fenwick, Peter Gariaev, Valerie Hunt, Ezio Insinna, David Lorimer, Hugh MacPherson, Robert Morris, Richard Obousy, Marcel Odier, Beverly Rubik, Rupert Sheldrake, Dennis Stillings, William Tiller, Marcel Truzzi, Dieter Vaitl, Harald Walach, Hans Wendt and Tom Williamson.

Although scores of books and papers contributed in some way to my thoughts and conclusions, I am indebted to Dean Radin's *The Conscious Universe: The Scientific Truth of Psychic Phenomena* (New York: HarperEdge, 1997) and Richard Broughton's *Parapsychology: The Controversial Science* (New York: Ballantine, 1991) for their compilation of evidence for psychic phenomena; Larry Dossey, whose various books were highly useful for evidence of spiritual healing; and Ervin Laszlo, for his fascinating theories of the vacuum in *The Interconnected Universe: Conceptual Foundations of Transdisciplinary Unified Theory* (Singapore: World Scientific, 1995).

I owe a special debt of gratitude to the team at HarperCollins, particularly my editors, Larry Ashmead and Krista Stroever, for their sage advice and courage in backing this project. I am especially grateful to Andrew Coleman, for his painstaking subediting of the manuscript. I am also indebted to my team at *What Doctors Don't Tell You* for their support. Julie

McLean and Sharyn Wong in particular offered vital aid at the eleventh hour, and Kathy Mingo's unfailing assistance enabled me to juggle home and work.

I owe a special thanks to Peter Robinson, my UK agent, and Daniel Benor, my international agent, for taking up the project with such enthusiasm. I should also particularly like to thank my agent in America, Russell Galen, whose dedication and unflagging belief in this project has been nothing short of astonishing.

Special mention must be made of my children, Caitlin and Anya, through whom I daily experience The Field firsthand. As ever, this book owes its largest debt to my husband Bryan Hubbard, for helping me to understand the true meaning of this book and also the true meaning of interconnection.

CONTENTS

PROLOGUE

The Coming Revolution

WE ARE POISED ON the brink of a revolution – a revolution as daring and profound as Einstein's discovery of relativity. At the very frontier of science new ideas are emerging that challenge everything we believe about how our world works and how we define ourselves. Discoveries are being made that prove what religion has always espoused: that human beings are far more extraordinary than an assemblage of flesh and bones. At its most fundamental, this new science answers questions that have perplexed scientists for hundreds of years. At its most profound, this is a science of the miraculous.

For a number of decades respected scientists in a variety of disciplines all over the world have been carrying out well-designed experiments whose results fly in the face of current biology and physics. Together, these studies offer us copious information about the central organizing force governing our bodies and the rest of the cosmos.

What they have discovered is nothing less than astonishing. At our most elemental, we are not a chemical reaction, but an energetic charge. Human beings and all living things are a coalescence of energy in a field of energy connected to every other thing in the world. This pulsating energy field is the central engine of our being and our consciousness, the alpha and the omega of our existence.

There is no 'me' and 'not-me' duality to our bodies in relation to the universe, but one underlying energy field. This field is responsible for our mind's highest functions, the information source guiding the growth of our bodies. It is our brain, our heart, our memory – indeed, a blueprint of the world for all time. The field is the force, rather than germs or genes, that finally determines whether we are healthy or ill, the force which must be tapped in order to heal. We are attached and engaged, indivisible from our world, and our only fundamental truth is our relationship with it. 'The field,' as Einstein once succinctly put it, 'is the only reality.'[1]

Up until the present, biology and physics have been handmaidens of views espoused by Isaac Newton, the father of modern physics. Everything we believe about our world and our place within it takes its lead from ideas that were formulated in the seventeenth century, but still form the backbone of modern science – theories that present all the elements of the universe as isolated from each other, divisible and wholly self-contained.

These, at their essence, created a world view of separateness. Newton described a material world in which individual particles of matter followed certain laws of motion through space and time – the universe as machine. Before Newton formulated his laws of motion, French philosopher René Descartes had come up with what was then a revolutionary notion, that we – represented by our minds – were separate from this lifeless inert matter of our bodies, which were just another type of well-oiled machine. The world was composed of a load of little discrete objects, which behaved predictably. The most separate of these was the human being. We sat outside this universe, looking in. Even our bodies were somehow separate and *other* from the real us, the conscious minds doing the observing.

The Newtonian world might have been law-abiding, but ultimately it was a lonely, desolate place. The world carried on, one vast gearbox, whether we were present or not. With a few deft moves, Newton and Descartes had plucked God and life from the world of matter, and us and our consciousness from the center of our world. They ripped the heart and soul out of the universe, leaving in its wake a lifeless collection of interlocking parts. Most important of all, as Danah Zohar observed in *The Quantum Self*, 'Newton's vision tore us out from the fabric of the universe.'[2]

Our self-image grew even bleaker with the work of Charles Darwin. His theory of evolution – tweaked slightly now by the neo-Darwinists – is of a life that is random, predatory, purposeless and solitary. Be the best or don't survive. You are no more than an evolutionary accident. The vast checkerboard biological heritage of your ancestors is stripped down to one central facet: survival. Eat or be eaten. The essence of your humanity is a genetic terrorist, efficiently disposing of any weaker links. Life is not about sharing and interdependence. Life is about winning, getting there first. And if you do manage to survive, you are on your own at the top of the evolutionary tree.

These paradigms – the world as machine, man as survival machine – have led to a technological mastery of the universe, but little real knowledge of any central importance to us. On a spiritual and metaphysical level, they have led to the most desperate and brutal sense of isolation. They also have got us no closer to understanding the most fundamental mysteries of our own being: how we think, how life begins, why we get ill, how a single cell turns into a fully formed person, and even what happens to human consciousness when we die.

We remain reluctant apostles of these views of the world as mechanized and separate, even if this isn't part of our ordinary experience. Many of us seek refuge from what we see as the harsh and nihilistic fact of our existence in religion, which may offer some succour in its ideals of unity, community and purpose, but through a view of the world that contradicts the view espoused by science. Anyone seeking a spiritual life has had to wrestle with these opposing world views and fruitlessly try to reconcile the two.

This world of the separate should have been laid waste once and for all by the discovery of quantum physics in the early part of the twentieth century. As the pioneers of quantum physics peered into the very heart of matter, they were astounded by what they saw. The tiniest bits of matter weren't even matter, as we know it, not even a set *something*, but sometimes one thing, sometimes something quite different. And even stranger, they were often many possible things all at the same time. But most significantly, these subatomic particles had no meaning in isolation, but only in relationship with everything else. At its most elemental, matter couldn't be chopped up into self-contained little units, but was completely indivisible. You could only understand the universe as a dynamic web of interconnection. Things once in contact remained always in contact through all space and all time. Indeed, time and space themselves appeared to be arbitrary constructs, no longer applicable at this level of the world. Time and space as we know them did not, in fact, exist. All that appeared, as far as the eye could see, was one long landscape of the here and now.

The pioneers of quantum physics – Erwin Schrödinger, Werner Heisenberg, Niels Bohr and Wolfgang Pauli – had some inkling of the metaphysical territory they had trespassed into. If electrons were connected everywhere at once, this implied something profound about the nature of the world at large. They turned to classic philosophical texts in their attempt to grasp the deeper truth about the strange subatomic world they were observing. Pauli examined psychoanalysis and archetypes and the Qabbalah; Bohr, the Tao and Chinese philosophy; Schrödinger, Hindu philosophy; and Heisenberg, the Platonic theory of ancient Greece.[3] Nevertheless, a coherent theory of the spiritual implications of quantum physics remained beyond their grasp. Niels Bohr hung a sign on his door saying 'Philosophers keep out. Work in progress.'

There was other, quite practical, unfinished business with quantum theory. Bohr and his colleagues only got so far in their experiments and understanding. The experiments they'd conducted demonstrating these quantum effects had occurred in the laboratory, with non-living subatomic particles. From there, scientists in their wake naturally assumed that this strange quantum world only existed in the world of dead matter. Anything alive still operated according to the laws of Newton and Descartes, a view that has informed all of modern medicine and biology. Even biochemistry depends upon Newtonian force and collision to work.

And what of us? Suddenly, we had grown central to every physical process, but no one had fully acknowledged this. The quantum pioneers had discovered that our involvement with matter was crucial. Subatomic particles existed in all possible states until disturbed by us – by observing or measuring – at which point, they'd settle down, at long last, into something real. Our observation – our human consciousness – was utterly central to this process of subatomic flux actually becoming some set thing, but we weren't in any of the mathematics of Heisenberg or Schrödinger. They realized that we were somehow key, but they didn't know how to include us. As far as science was concerned, we were still on the outside looking in.

All the loose strands of quantum physics were never tied up into a coherent theory, and quantum physics got reduced to an extremely successful tool of technology, vital for making bombs and modern electronics. The philosophical implications were forgotten, and all that remained were its practical advantages. The rank and file of today's physicists were willing to accept the bizarre nature of the quantum world at face value because the mathematics, such as the Schrödinger equation, works so well, but shook their heads at the counter-intuitiveness of it all.[4] How could electrons be in touch with everything at once? How could an electron not be a set single thing until it is examined or measured? How, in fact, could anything be concrete in the world, if it was a will o' the wisp once you started looking closer at it?

Their answer was to say that there was a single truth for anything small and another truth for something much bigger, one truth for things that were alive, another for things that weren't, and to accept these apparent contradictions just as one might accept a basic axiom of Newton's. These were the rules of the world and they should just be taken at face value. The math works, and that's all that counts.

A small band of scientists dotted around the globe was not satisfied to simply carry on with quantum physics by rote. They required a better answer to many of the large questions that had been left unanswered. In their investigations and experimentation, they picked up where the pioneers of quantum physics had left off, and they began probing deeper.

Several thought again about a few equations that had always been subtracted out in quantum physics. These equations stood for the Zero Point Field – an ocean of microscopic vibrations in the space between things. If the Zero Point Field were included in our conception of the most fundamental nature of matter, they realized, the very underpinning of our universe was a heaving sea of energy – one vast quantum field. If this were true, everything would be connected to everything else like some invisible web.

They also discovered that we were made of the same basic material. On our most fundamental level, living beings, including human beings, were packets of quantum energy constantly exchanging information with this inexhaustible energy sea. Living things emitted a weak radiation, and this was the most crucial aspect of biological processes. Information about all aspects of life, from cellular communication to the vast array of controls of DNA, was relayed through an information exchange on the quantum level. Even our minds, that *other* supposedly so outside of the laws of matter, operated according to quantum processes. Thinking, feeling – every higher cognitive function – had to do with quantum information pulsing simultaneously through our brains and body. Human perception occurred because of interactions between the subatomic particles of our brains and the quantum energy sea. We literally resonated with our world.

Their discoveries were extraordinary and heretical. In a stroke, they had challenged many of the most basic laws of biology and physics. What they may have uncovered was no less than the key to all information processing and exchange in our world, from the communication between cells to perception of the world at large. They'd come up with answers to some of the most profound questions in biology about human morphology and living consciousness. Here, in so-called 'dead' space, possibly lay the very key to life itself.

Most fundamentally, they had provided evidence that all of us connect with each other and the world at the very undercoat of our being. Through scientific experiment they'd demonstrated that there may be such a thing as a life force flowing through the universe – what has variously been called collective consciousness or, as theologians have termed it, the Holy Spirit. They provided a plausible explanation of all those areas that over the centuries mankind has had faith in but no solid evidence of or adequate accounting for, from the effectiveness of alternative medicine and even prayer to life after death. They offered us, in a sense, a science of religion.

Unlike the world view of Newton or Darwin, theirs was a vision that was life-enhancing. These were ideas that could empower us, with their implications of order and control. We were not simply accidents of nature. There was purpose and unity to our world and our place within it, and we had an important say in it. What we did and thought mattered – indeed, was critical in creating our world. Human beings were no longer separate from each other. It was no longer us and them. We were no longer at the periphery of our universe – on the outside looking in. We could take our rightful place, back in the center of our world.

These ideas were the stuff of treason. In many cases, these scientists have had to fight a rearguard action against an entrenched and hostile establishment. Their investigations have gone on for thirty years, largely unacknowledged or suppressed, but not because of the quality of the work. The scientists, all from credible top-ranking institutions – Princeton University, Stanford University, top institutions in Germany and France – have produced impeccable experimentation. Nevertheless, their experiments have attacked a number of tenets held to be sacred and at the very heart of modern science. They did not fit the prevailing scientific view of the world – the world as machine. Acknowledging these new ideas would require scrapping much of what modern science believes in and, in a sense, starting over from scratch. The old guard was having none of it. It did not fit the world view and so it must be wrong.

Nevertheless, it is too late. The revolution is unstoppable. The scientists who have been highlighted in *The Field* are merely a few of the pioneers, a small representation of a larger movement.[5] Many others are right behind them, challenging, experimenting, modifying their views, engaged in the work that all true explorers engage in. Rather than dismissing this information as not fitting in with the scientific view of the world, orthodox science will have to begin adapting its world view to suit. It is time to relegate Newton and Descartes to their proper places, as prophets of a historical view that has now been surpassed. Science can only be a process of understanding our world and ourselves, rather than a fixed set of rules for all time, and with the ushering in of the new, the old must often be discarded.

The Field is the story of this revolution in the making. Like many revolutions, it began with small pockets of rebellion, which gathered individual strength and momentum – a breakthrough in one area, a discovery somewhere else – rather than one large, unified movement of reform. Although aware of each other's work, these are men and women in the laboratory, who often dislike venturing beyond experimentation to examine the full implications of their findings or don't always have the time necessary to place them in context with other scientific evidence coming to light. Each scientist has been on a voyage of discovery, and each has discovered a bucket of earth, but no one has been bold enough to declare it a continent.

The Field represents one of the first attempts to synthesize this disparate research into a cohesive whole. In the process, it also provides a scientific validation of areas which have largely been the domain of religion, mysticism, alternative medicine or New Age speculation.

Although all of the material in this book is grounded in the hard fact of scientific experimentation, at times, with the help of the scientists concerned, I've had to engage in speculation as to how all this fits together. Consequently, I must stress that this theory is, as Princeton Dean Emeritus Robert Jahn is fond of saying, a work in progress. In a few instances, some of the scientific evidence presented in *The Field* has not yet been reproduced by independent groups. As with all new ideas, *The Field* has to be seen as an early attempt to put individual findings into a coherent model, portions of which are bound to be refined in future.

It is also wise to keep in mind the well-known dictum that a right idea can never get definitively proven. The best that science can ever hope to achieve is to disprove wrong ideas. There have been many attempts to discredit the new ideas elaborated in this book by scientists with good credentials and testing methods, but thus far, no one has been successful. Until they are disproven or refined, the findings of these scientists stand as valid.

This book is intended for a lay audience, and in order to make quite complicated notions comprehensible, I've often had to reach for metaphors which represent only a crude approximation of the truth. At times, the radical new ideas presented in this book will require patience, and I cannot promise that this will always be an easy read. A number of notions are quite difficult for the Newtonians and Cartesians among us, accustomed as we are to thinking of everything in the world as separate and inviolate.

It is also important to stress that none of this is my discovery. I am not a scientist. I am only the reporter and occasionally the interpreter. The plaudits go to the largely unknown men and women in the laboratory who have unearthed and grasped the extraordinary in the course of the everyday. Often without their even fully comprehending it, their work transformed into a quest for the physics of the impossible.

Lynne McTaggart
London, July 2001

Part 1

The Resonating Universe

Now I know we're not in Kansas.

Dorothy, *The Wizard of Oz*

CHAPTER ONE

Light in the Darkness

PERHAPS WHAT HAPPENED TO Ed Mitchell was due to the lack of gravity, or maybe to the fact that all his senses had been disoriented. He had been on his way home, which at the moment was approximately 250,000 miles away, somewhere on the surface of the clouded azure and white crescent appearing intermittently through the triangular window of the command module of the *Apollo 14*.[1]

Two days before, he had become the sixth man to land on the moon. The trip had been a triumph: the first lunar landing to carry out scientific investigations. The 94 pounds of rock and soil samples in the hold attested to that. Although he and his commander, Alan Shepard, hadn't reached the summit of the 750-foot-high ancient Cone Crater, the rest of the items on the meticulous schedule taped to their wrists, detailing virtually every minute of their two-day journey, had been methodically ticked off.

What they hadn't fully accounted for was the effect of this uninhabited world, low in gravity, devoid of the diluting effect of atmosphere, on the senses. Without signposts such as trees or telephone wires, or indeed anything other than the *Antares*, the gold insect-like lunar module, on the full sweep of the dust-grey landscape, all perceptions of space, scale, distance or depth were horribly distorted; Ed had been shocked to discover that any points of navigation which had been carefully noted on high-resolution photographs were at least double the distance expected. It was as though he and Alan had shrunk during space travel and what from home had appeared to be tiny humps and ridges on the moon's surface had suddenly swollen to heights of six feet or more. And yet if they felt diminished in size, they were also lighter than ever. He'd experienced an odd lightness of being, from the weak gravitational pull, and despite the weight and bulk of his ungainly spacesuit, felt buoyed at every step.

There had also been the distorting effect of the sun, pure and unadulterated in this airless world. In the blinding sunlight, even in the relatively cool morning, before the highs that might reach 270° F, craters, landmarks, soil and the earth – even the sky itself – all stood out in absolute clarity. For a mind accustomed to the soft filter of atmosphere, the sharp shadows, the changeable colors of the slate-grey soil all conspired to play tricks on the eye. Unknowingly he and Alan had been only 61 feet from Cone Crater's edge, about 10 seconds away, when they turned back, convinced that they wouldn't reach it in time – a failure that would bitterly disappoint Ed, who'd longed to stare into that 1100-foot diameter hole in the midst of the lunar uplands. Their eyes didn't know how to interpret this hyperstate of vision. Nothing lived, but also nothing was hidden from view, and everything lacked subtlety. Every sight overwhelmed the eye with brilliant contrasts and shadows. He was seeing, in a sense, more clearly and less clearly than he ever had.

During the relentless activity of their schedule, there had been little time for reflection or wonder, or for any thoughts of a larger purpose to the trip. They had gone farther in the universe than any man before them, and yet, weighed down by the knowledge that they were costing the American taxpayers $200,000 a minute, they felt compelled to keep their eyes on the clock, ticking off the details of what Houston had planned in their packed schedule. Only after the lunar module had reconnected with the command module and begun the two-day journey back to earth could Ed pull off his spacesuit, now filthy with lunar soil, sit back in his long johns and try to put his frustration and his jumble of thoughts into some sort of order.

The *Kittyhawk* was slowly rotating, like a chicken on a spit, in order to balance the thermal effect on each side of the spacecraft; and in its slow revolution, earth was intermittently framed through the window as a tiny crescent in an all-engulfing night of stars. From this perspective, as the earth traded places in and out of view with the rest of the solar system, sky didn't exist only above the astronauts, as we ordinarily view it, but as an all-encompassing entity that cradled the earth from all sides.

It was then, while staring out of the window, that Ed experienced the strangest feeling he would ever have: a feeling of *connectedness*, as if all the planets and all the people of all time were attached by some invisible web. He could hardly breathe from the majesty of the moment. Although he continued to turn knobs and press buttons, he felt distanced from his body, as though someone else were doing the navigating.

There seemed to be an enormous force field here, connecting all people, their intentions and thoughts, and every animate and inanimate form of matter for all time. Anything he did or thought would influence the rest of the cosmos, and every occurrence in the cosmos would have a similar effect on him. Time was just an artificial construct. Everything he'd been taught about the universe and the separateness of people and things felt wrong. There were no accidents or individual intentions. The natural intelligence that had gone on for billions of years, that had forged the very molecules of his being, was also responsible for his own present journey. This wasn't something he was simply comprehending in his mind, but an overwhelmingly visceral feeling, as though he were physically extending out of the window to the very furthest reaches of the cosmos.

He hadn't seen the face of God. It didn't feel like a standard religious experience so much as a blinding epiphany of meaning – what the Eastern religions often term an 'ecstasy of unity'. It was as though in a single instant Ed Mitchell had discovered and felt The Force.

He stole a glance at Alan and Stu Roosa, the other astronaut on the *Apollo 14* mission, to see if they were experiencing anything remotely similar. There had been a moment when they'd first stepped off the *Antares* and into the plains of Fra Mauro, a highland region of the moon, when Alan, a veteran of the first American space launch, ordinarily so hard-boiled, with little time for this kind of mystical mumbo-jumbo, strained in his bulky spacesuit to look up above him and wept at the sight of the earth, so impossibly beautiful in the airless sky. But now Alan and Stu appeared to be automatically going about their business, and so he was afraid to say anything about what was beginning to feel like his own ultimate moment of truth.

He'd always been a bit of the odd man out in the space program and certainly, at 41, although younger than Shepard, he was one of the senior members of *Apollo*. Oh, he looked and acted the part all right, with his sandy-haired, broad-faced, Midwestern looks and the languid drawl of a commercial airline pilot. But to the others, he was a bit of an intellectual: the only one among them with both a PhD and test-pilot credentials. The way he'd entered the space program had been decidedly left field. Getting his doctorate in astrophysics from MIT was the way he thought he'd be indispensable – that's how deliberately he'd plotted his path toward NASA – and only afterward did it occur to him to boost the flying time he'd gained overseas to qualify. Nevertheless, Ed was no slouch when it came to flying. Like all the other fellows, he'd put in his time at Chuck Yeager's flying circus in the Mojave Desert, getting airplanes to do things they'd never been designed to do. At one point, he'd even been their instructor. But he liked to think of himself as not so much a test pilot as an explorer: a kind of modern-day seeker after truths. His own attraction toward science constantly wrestled with the fierce Baptist fundamentalism of his youth. It seemed no accident that he'd grown up in Roswell, New Mexico, where the first alien sightings supposedly had occurred – just a mile down the road from the home of Robert Goddard, the father of American rocket science, and just a few miles across the mountains from the first testings of the atomic bomb. Science and spirituality coexisted in him, jockeying for position, but he yearned for them to somehow shake hands and make peace.

There was something else he'd kept from them. Later that evening, as Alan and Stu slept in their hammocks, Ed silently pulled out what had been an ongoing experiment during the whole of his journey to and from the moon. Lately, he'd been dabbling in experiments in consciousness and extrasensory perception, spending time studying the work of Dr Joseph B. Rhine, a biologist who'd conducted many experiments on the extrasensory nature of human consciousness. Two of his newest friends were doctors who'd been conducting credible experiments on the nature of consciousness. Together they'd realized that Ed's journey to the moon presented them with a unique opportunity to test whether human telepathy could be achieved at greater distances than it had in Dr Rhine's laboratory. Here was a once-in-a-lifetime chance to see if these sorts of communications could stretch well beyond any distances possible on earth.

Forty-five minutes past the start of the sleep period, as he had done in the two days traveling to the moon, Ed pulled out a small flashlight and, on the paper on his clipboard, randomly copied numbers, each of which stood for one of Dr Rhine's famous Zener symbols – square, circle, cross, star, and pair of wavy lines. He'd then concentrated intensely on them, methodically, one by one, attempting to 'transmit' his choices to his colleagues back home. As excited as he was about it, he kept the experiment to himself. Once he'd tried to have a discussion with Alan about the nature of consciousness, but he wasn't really close to his boss and it wasn't the sort of issue that burned in the others like it did in him. Some of the astronauts had thought about God while they were out in space, and everybody in the entire space program knew they were looking for something new about the way the universe worked. But if Alan and Stu had known that he was trying to transmit his thoughts to people on earth, they would have thought him more of an oddball than they did already.

Ed finished the night's experiment and would do another one the following evening. But after what had happened to him earlier, it hardly seemed necessary any more; he now had his own inner conviction that it was true. Human minds were connected to each other, just as they were connected to everything else in this world and every other world. The intuitive in him accepted that, but for the scientist in him it wasn't enough. For the next 25 years he'd be looking to science to explain to him what on earth it was that had happened to him out there.

Edgar Mitchell got home safely. No other physical exploration on earth could possibly compare with going to the moon. Within the next two years he left NASA when the last three lunar flights were canceled for lack of funds, and that was when the real journey began. Exploring inner space would prove infinitely longer and more difficult than landing on the moon or searching out Cone Crater.

His little experiment with ESP was successful, suggesting that some form of communication defying all logic had taken place. Ed hadn't been able to do all six experiments as planned and it took some time to match the four he'd managed with the six sessions of guessing which had been conducted on earth. But when the four sets of data Ed had amassed during the nine-day journey were finally matched with those of his six colleagues on earth, the correspondence between them was shown to be significant, with a one in 3000 probability that this was due to chance.[2] These results were in line with thousands of similar experiments conducted on earth by Rhine and his colleagues over the years.

Edgar Mitchell's lightning-bolt experience while in space had left hairline cracks in a great number of his belief systems. But what bothered Ed most about the experience he had in outer space was the current scientific explanation for biology and particularly consciousness, which now seemed impossibly reductive. Despite what he'd learned in quantum physics about the nature of the universe, during his years at MIT, it seemed that biology remained mired in a 400-year-old view of the world. The current biological model still seemed to be based on a classical Newtonian view of matter and energy, of solid, separate bodies moving predictably in empty space, and a Cartesian view of the body as separate from the soul, or mind. Nothing in this model could accurately reflect the true complexity of a human being, its relation to its world or, most particularly, its consciousness; human beings and their parts were still treated, for all intents and purposes, as machinery.

Most biological explanations of the great mysteries of living things attempt to understand the whole by breaking it down into ever more microscopic parts. Bodies supposedly take the shape they do because of genetic imprinting, protein synthesis and blind mutation. Consciousness resided, according to the neuroscientists of the day, in the cerebral cortex – the result of a simple mix between chemicals and brain cells. Chemicals were responsible for the television set playing out in our brain, and chemicals were responsible for the 'it' that is doing the viewing.[3] We know the world because of the intricacies of our own machinery. Modern biology does not believe in a world that is ultimately indivisible.

In his own work on quantum physics at MIT, Ed Mitchell had learned that at the subatomic level, the Newtonian, or classical, view – that everything works in a comfortably predictable manner – had long been replaced by messier and indeterminate quantum theories, which suggest that the universe and the way it works are not quite as tidy as scientists used to think.

Matter at its most fundamental level could not be divided into independently existing units or even be fully described. Subatomic particles weren't solid little objects like billiard balls, but vibrating and indeterminate packets of energy that could not be precisely quantified or understood in themselves. Instead, they were schizophrenic, sometimes behaving as particles – a set thing confined to a small space – and sometimes like a wave – a vibrating and more diffuse thing spread out over a large region of space and time – and sometimes like both a wave and a particle at the same time. Quantum particles were also omnipresent. For instance, when transiting from one energy state to another, electrons seemed to be testing out all possible new orbits at once, like a property buyer attempting to live in every house on the block *at the same instant* before choosing which one to finally settle in. And nothing was certain. There were no definite locations, but only a likelihood that an electron, say, might be at a certain place, no set occurrence but only a probability that it might happen. At this level of reality, nothing was guaranteed; scientists had to be content with only being able to bet on the odds. The best that ever could be calculated was probability – the likelihood, when you take a certain measurement, that you will get a certain result a certain percentage of the time. Cause-and-effect relationships no longer held at the subatomic level. Stable-looking atoms might suddenly, without apparent cause, experience some internal disruption; electrons, for no reason, elect to transit from one energy state to another. Once you peered closer and closer at matter, it wasn't even matter, not a single solid thing you could touch or describe, but a host of tentative selves, all being paraded around at the same time. Rather than a universe of static certainty, at the most fundamental level of matter, the world and its relationships were uncertain and unpredictable, a state of pure potential, of infinite possibility.

Scientists did allow for a universal connectedness in the universe, but only in the quantum world: which was to say, the realm of the inanimate and not the living. Quantum physicists had discovered a strange property in the subatomic world called 'nonlocality'. This refers to the ability of a quantum entity such as an individual electron to influence another quantum particle instantaneously over any distance despite there being no exchange of force or energy. It suggested that quantum particles once in contact retain a connection even when separated, so that the actions of one will always influence the other, no matter how far they get separated. Albert Einstein disparaged this 'spooky action at a distance', and it was one of the major reasons he so distrusted quantum mechanics, but it has been decisively verified by a number of physicists since 1982.[4]

Nonlocality shattered the very foundations of physics. Matter could no longer be considered separate. Actions did not have to have an observable cause over an observable space. Einstein's most fundamental axiom wasn't correct: at a certain level of matter, things could travel faster than the speed of light. Subatomic particles had no meaning in isolation but could only be understood in their relationships. The world, at its most basic, existed as a complex web of interdependent relationships, forever indivisible.

Perhaps the most essential ingredient of this interconnected universe was the living consciousness that observed it. In classical physics, the experimenter was considered a separate entity, a silent observer behind glass, attempting to understand a universe that carried on, whether he or she was observing it or not. In quantum physics, however, it was discovered, the state of all possibilities of any quantum particle collapsed into a set entity as soon as it was observed or a measurement taken. To explain these strange events, quantum physicists had postulated that a participatory relationship existed between observer and observed – these particles could only be considered as 'probably' existing in space and time until they were 'perturbed', and the act of observing and measuring them forced them into a set state – an act akin to solidifying Jell-O. This astounding observation also had shattering implications about the nature of reality. It suggested that the consciousness of the observer brought the observed object into being. Nothing in the universe existed as an actual 'thing' independently of our perception of it. Every minute of every day we were creating our world.

It seemed a central paradox to Ed that physicists would have you believe that sticks and stones have a different set of physical rules from the atomic particles within them, that there should be one rule for the tiny and one for the large, one rule for the living, another for the inert. Classical laws were undoubtedly useful for fundamental properties of motion, in describing how skeletons hold us up or how our lungs breathe, our hearts pump, our muscles carry heavy weights. And many of the body's basic processes – eating, digestion, sleeping, sexual function – are indeed governed by physical laws.

But classical physics or biology could not account for such fundamental issues as how we can think in the first place; why cells organize as they do; how many molecular processes proceed virtually instantaneously; why arms develop as arms and legs as legs, even though they have the same genes and proteins; why we get cancer; how this machine of ours can miraculously heal itself; and even what knowing is – how it is that we know what we know. Scientists might understand in minute detail the screws, bolts, joints and various wheels, but nothing about the force that powers the engine. They might treat the smallest mechanics of the body but still they appeared ignorant of the most fundamental mysteries of life.

If it were true that the laws of quantum mechanics also apply to the world at large, and not just the subatomic world, and to biology and not just the world of matter, then the entire paradigm for biological science was flawed or incomplete. Just as Newton's theories had eventually been improved upon by the quantum theorists, perhaps Heisenberg and Einstein themselves had been wrong, or at least only partially right. If quantum theory were applied to biology on a larger scale, we would be viewed more as a complex network of energy fields in some sort of dynamic interplay with our chemical cellular systems. The world would exist as a matrix of indivisible interrelation, just as Ed had experienced it in outer space. What was so evidently missing from standard biology was an explanation for the organizing principle – for human consciousness.

Ed began devouring books about religious experiences, Eastern thought, and the little scientific evidence that existed on the nature of consciousness. He launched early studies with a number of scientists in

Stanford; he set up the Institute of Noetic Sciences, a non-profit organization whose role was to fund this type of research; he began amassing scientific studies of consciousness into a book. Before long, it was all he could think of and talk about, and what had turned into an obsession tore his marriage apart.

Edgar's work may not have lit a revolutionary fire, but he certainly stoked it. In prestigious universities around the world tiny pockets of quiet rebellion were sprouting up against the world view of Newton and Darwin, the dualism in physics and the current view of human perception. During his search, Ed began making contact with scientists with impressive credentials at many of the big reputable universities – Yale, Stanford, Berkeley, Princeton, the University of Edinburgh – who were coming up with discoveries that just didn't fit.

Unlike Edgar, these scientists hadn't undergone an epiphany to arrive at a new world view. It was simply that in the course of their work they'd come across scientific results which were square pegs to the round hole of established scientific theory, and much as they might try to jam them into place – and in many cases, the scientists wished, indeed willed, them to fit – they would stubbornly resist. Most of the scientists had arrived at their conclusions accidentally, and, as if they'd landed at the wrong railway station, once they'd got there, they figured that there was no other possibility but to get out and explore the new terrain. To be a true explorer is to carry on your exploration even if it takes you to a place you didn't particularly plan to go to.

The most important quality common to all these researchers was a simple willingness to suspend disbelief and remain open to true discovery, even if it meant challenging the existing order of things, alienating colleagues or opening themselves up to censure and professional ruin. To be a revolutionary in science today is to flirt with professional suicide. Much as the field purports to encourage experimental freedom, the entire structure of science, with its highly competitive grant system, coupled with the publishing and peer review system, largely depends upon individuals conforming to the accepted scientific world view. The system tends to encourage professionals to carry out experimentation whose purpose is primarily to confirm the existing view of things, or to further develop technology for industry, rather than to serve up true innovation.[5]

Everyone working on these experiments had the sense that they were on the verge of something that was going to transform everything we understood about reality and human beings, but at the time they were simply frontier scientists operating without a compass. A number of scientists working independently had come up with a single bit of the puzzle and were frightened to compare notes. There was no common language because what they were discovering appeared to *defy* language.

Nevertheless, as Mitchell made contact with them, their separate work began to coalesce into an alternative theory of evolution, human consciousness and the dynamics of all living things. It offered the best prospect for a unified view of the world based on actual experimentation and mathematical equations, and not simply theory. Ed's major role was making introductions, funding some of the research and, through his willingness to use his celebrity status as a national hero to make this work public, convincing them that they were not alone.

All the work converged on a single point – that the self had a field of influence on the world and vice versa. There was one other point of common agreement: all the experiments being carried out drove a stake into the very heart of existing scientific theory.

CHAPTER TWO

The Sea of Light

Bill church was out of gas. Ordinarily, this would not be a situation that could ruin an entire day. But in 1973, in the grip of America's first oil crisis, getting your car filled up with gas depended upon two things: the day of the week and the last number of your license plate. Those whose plates ended in an odd number were allowed to fill up on Mondays, Wednesdays or Fridays; even numbers on Tuesdays, Thursdays and Saturdays, with Sunday a gas-free day of rest. Bill had an odd number and the day was Tuesday. That meant that no matter where he had to go, no matter how important his meetings, he was stuck at home, held hostage by a few Middle Eastern potentates and OPEC. Even if his license plate number matched the day of the week, it still could take up to two hours waiting in lines that zigzagged around corners many blocks away. That is, if he could find a gas station that was still open.

Two years before, there had been plenty of fuel to send Edgar Mitchell to the moon and back. Now half the country's gas stations had gone out of business. President Nixon had recently addressed the nation, urging all Americans to turn down their thermostats, form car pools and use no more than 10 gallons a week. Businesses were asked to halve the lighting in work areas and to turn down lights in halls and storage areas. Washington would set the example by keeping the national Christmas tree on the White House front lawn turned off. The nation, fat and complacent, used to consuming energy like so many cheeseburgers, was in shock, forced, for the first time, to go on a diet. There was talk of rationing books being printed. Five years later Jimmy Carter would term it the 'moral equivalent of war', and it felt that way to most middle-aged Americans, who hadn't had to ration gas since the Second World War.

Bill stormed back inside and got on the phone to Hal Puthoff to complain. Hal, a laser physicist, often acted as Bill's scientific alter-ego. 'There has got to be a better way,' Bill shouted frustratedly.

Hal agreed that it was time to start looking for some alternatives to fossil fuel to drive transportation – something besides coal, wood or nuclear power.

'But what else is there?' said Bill.

Hal ticked off a litany of current possibilities. There was photovoltaics (using solar cells), or fuel cells, or water batteries (an attempt to convert the hydrogen from water into electricity in the cell). There was wind, or waste products, or even methane. But none of these, even the more exotic among them, were turning out to be robust or realistic.

Bill and Hal agreed that what was really needed was an entirely new source: a cheap, endless, perhaps as yet undiscovered, supply of energy. Their conversations often veered off in this kind of speculative direction. Hal, in the main, liked cutting-edge technology – the more futuristic, the better. He was more an inventor than your ordinary physicist, and at 35 already had a patent on a tuneable infrared laser. Hal was largely self-made and had put himself through school after his father died when he was in his early teens. He'd graduated from the University of Florida in 1958, the year after *Sputnik 1* went up, but he'd come of age during the Kennedy administration. Like many young men of his generation, he'd taken to heart Kennedy's central metaphor of the US embarking on a new frontier. Through the years and even after the space program had fallen away due to lack of interest as well as lack of funding, Hal would retain a humble idealism about his work and the central role science played in the future of mankind. Hal firmly believed that science drove civilization. He was a small, sturdy man with a passing resemblance to Mickey Rooney and a sweep of thick chestnut hair, whose seething inner life of lateral thought and what-if possibility hid behind a phlegmatic and unassuming exterior. At first glance, he hardly looked the part of the frontier scientist. Nevertheless, it was Hal's sincere view that frontier work was vital for the future of the planet, to provide inspiration for teaching and for economic growth. He also liked getting out of the laboratory, trying to apply physics to solutions in real life.

Bill Church might be a successful businessman, but he shared much of Hal's idealism about science improving civilization. He was a modest Medici to Hal's Da Vinci. Bill had cut his own career in science short when he was drafted to run the family business, Church's Fried Chicken, the Texan answer to Kentucky Fried Chicken. He'd spent 10 years at it and recently he'd taken Church's to the market. He'd made his money and now he was in the mood to return to his youthful aspirations – but with no education, he'd had to do it by proxy. In Hal he'd found his perfect counterpart – a gifted physicist willing to pursue areas that ordinary scientists might dismiss out of hand. In September 1982, Bill would present Hal with a gold watch to mark their collaboration: 'To Glacier Genius from Snow,' it read. The idea was that Hal was the quiet innovator, tenacious and cool as a glacier, with Bill as 'Snow', throwing new challenges at him like a constant barrage of fine new powder.

'There is one giant reservoir of energy we haven't talked about,' Hal said. Every quantum physicist, he explained, is well aware of the Zero Point Field. Quantum mechanics had demonstrated that there is no such thing as a vacuum, or nothingness. What we tend to think of as a sheer void if all of space were emptied of matter and energy and you examined even the space between the stars is, in subatomic terms, a hive of activity.

The uncertainty principle developed by Werner Heisenberg, one of the chief architects of quantum theory, implies that no particle ever stays completely at rest but is constantly in motion due to a ground state field of energy constantly interacting with all subatomic matter. It means that the basic substructure of the universe is a sea of quantum fields that cannot be eliminated by any known laws of physics.

What we believe to be our stable, static universe is in fact a seething maelstrom of subatomic particles fleetingly popping in and out of existence. Although Heisenberg's principle most famously refers to the uncertainty attached to measuring the physical properties of the subatomic world, it also has another meaning: that we cannot know both the energy and the lifetime of a particle, so a subatomic event occurring within a tiny time frame involves an uncertain amount of energy. Largely because of Einstein's theories and his famous equation $E = mc^2$, relating energy to mass, all elementary particles interact with each other by exchanging energy through other quantum particles, which are believed to appear out of nowhere, combining and annihilating each other in less than an instant – 10^{-23} seconds, to be exact – causing random fluctuations of energy without any apparent cause. The fleeting particles generated during this brief moment are known as 'virtual particles'. They differ from real particles because they only exist during that exchange – the time of 'uncertainty' allowed by the uncertainty principle. Hal liked to think of this process as akin to the spray given off from a thundering waterfall.[1]

This subatomic tango, however brief, when added across the universe, gives rise to enormous energy, more than is contained in all the matter in all the world. Also referred to by physicists as 'the vacuum', the Zero Point Field was called 'zero' because fluctuations in the field are still detectable in temperatures of absolute zero, the lowest possible energy state, where all matter has been removed and nothing is supposedly left to make any motion. Zero-point energy was the energy present in the emptiest state of space at the lowest possible energy, out of which no more energy could be removed – the closest that motion of subatomic matter ever gets to zero.[2] But because of the uncertainty principle there will always be some residual jiggling due to virtual particle exchange. It had always been largely discounted because it is ever-present. In physics equations, most physicists would subtract troublesome zero-point energy away – a process called 'renormalization'.[3] Because zero-point energy was ever-present, the theory went, it didn't change anything. Because it didn't change anything, it didn't count.[4]

Hal had been interested in the Zero Point Field for a number of years, ever since he'd stumbled on the papers of Timothy Boyer of City University in New York in a physics library. Boyer had demonstrated that classical physics, allied with the existence of the ceaseless energy of the Zero Point Field, could explain many of the strange phenomena attributed to quantum theory.[5] If Boyer were to be believed, it meant that you didn't need two types of physics – the classical Newtonian kind and the quantum laws – to account for the properties of the universe. You could explain everything that happened in the quantum world with classical physics – so long as you took account of the Zero Point Field.

The more Hal thought about it, the more he became convinced that the Zero Point Field fulfilled all the criteria he was looking for: it was free; it was boundless; it didn't pollute anything. The Zero Point Field might just represent some vast unharnessed energy source. 'If you could just tap into this,' Hal said to Bill, 'you could even power spaceships.'

Bill loved the idea and offered to fund some exploratory research. It wasn't as though he hadn't funded crazier schemes of Hal's before. In a sense the timing was right for Hal. At 36, he was at a bit of a loose end. His first marriage had broken up, he'd just finished co-authoring what had become an important textbook on quantum electronics. He'd got his PhD in electrical engineering from Stanford just five years before, and had made his mark in lasers. When academia had proved tedious to him, he'd moved on, and was presently a laser researcher at Stanford Research Institute (SRI), a gigantic farmers' market of a research site, at the time affiliated with Stanford University. SRI stood like its own vast university of interlocking rectangles, squares and Zs of three-storey red-brick buildings hidden in a sleepy little corner of Menlo Park, sandwiched between St Patrick's seminary and the city of Spanish-tiled roofs representing Stanford University itself. At the time, SRI was the second largest think-tank in the world, where anyone could study virtually anything so long as they were able to get the funding for it.

Hal devoted several years to reading the scientific literature and doing some elementary calculations. He looked at other related aspects of the vacuum and general relativity in a more fundamental way. Hal, who tended toward the taciturn, attempted to keep himself within the confines of the purely intellectual, but occasionally he couldn't prevent his mind from giddily racing ahead. Even though these were early days, he knew he'd stumbled onto something of major significance for physics. This was an incredible breakthrough, possibly even a way to apply quantum physics to the world on a large scale, or perhaps it was a new science altogether. This was beyond lasers or anything else he had ever done. This felt, in its own modest way, a little like being Einstein and discovering relativity. Eventually, he realized just what it was that he had: he was on the verge of the discovery that the 'new' physics of the subatomic world might be wrong – or at least require some drastic revision.

Hal's discovery, in a sense, was not a discovery at all, but a situation that physicists have taken for granted since 1926 and discarded as immaterial. To the quantum physicist, it is an annoyance, to be subtracted away and discounted. To the religious or the mystic, it is science proving the miraculous. What quantum calculations show is that we and our universe live and breathe in what amounts to a sea of motion – a quantum sea of light. According to Heisenberg, who developed the uncertainty principle in 1927, it is impossible to know all the properties of a particle, such as its position and its momentum, at the same time because of what seem to be fluctuations inherent in nature. The energy level of any known particle can't be pinpointed because it is always changing. Part of this principle also stipulates that no subatomic particle can be brought completely to rest, but will always possess a tiny residual movement. Scientists have long known that these fluctuations account for the random noise of microwave receivers or electronic circuits, limiting the level to which signals can be amplified. Even fluorescent strip lighting relies on vacuum fluctuations to operate.

Imagine taking a charged subatomic particle and attaching it to a little frictionless spring (as physicists are fond of doing to work out their equations). It should bounce up and down for a while and then, at a temperature of absolute zero, stop moving. What physicists since Heisenberg have found is that the energy in the Zero Point Field keeps acting on the particle so that it never comes to rest but always keeps moving on the spring.[6]

Against the objections of his contemporaries, who believed in empty space, Aristotle was one of the first to argue that space was in fact a plenum (a background substructure filled with things). Then, in the middle of the nineteenth century, scientist Michael Faraday introduced the concept of a field in relation to electricity and magnetism, believing that the most important aspect of energy was not the source but the space around it, and the influence of one on the other through some force.[7] In his view, atoms weren't hard little billiard balls, but the most concentrated center of a force that would extend out in space.

A field is a matrix or medium which connects two or more points in space, usually via a force, like gravity or electromagnetism. The force is usually represented by ripples in the field, or waves. An electromagnetic field, to use but one example, is simply an electrical field and a magnetic field which intersect, sending out waves of energy at the speed of light. An electric and magnetic field forms around any electric charge (which is, most simply, a surplus or deficit of electrons). Both electrical and magnetic fields have two polarities (negative and positive) and both will cause any other charged object to be attracted or repelled, depending on whether the charges are opposite (one positive, the other negative) or the same (both positive or both negative). The field is considered that area of space where this charge and its effects can be detected.

The notion of an electromagnetic field is simply a convenient abstraction invented by scientists (and represented by lines of 'force', indicated by direction and shape) to try to make sense of the seemingly remarkable actions of electricity and magnetism and their ability to influence objects at a distance – and, technically, into infinity – with no detectable substance or matter in between. Simply put, a field is a region of influence. As one pair of researchers aptly described it: 'Every time you use your toaster, the fields around it perturb charged particles in the farthest galaxies ever so slightly.'[8]

James Clerk Maxwell first proposed that space was an ether of electromagnetic light, and this idea held sway until decisively disproved by a Polish-born physicist named Albert Michelson in 1881 (and six years later in collaboration with an American chemistry professor called Edward Morley) with a light experiment that showed that matter did not exist in a mass of ether.[9] Einstein himself believed space constituted a true void until his own ideas, eventually developed into his general theory of relativity, showed that space indeed held a plenum of activity. But it wasn't until 1911, with an experiment by Max Planck, one of the founding fathers of quantum theory, that physicists understood that empty space was bursting with activity.

In the quantum world, quantum fields are not mediated by forces but by exchange of energy, which is constantly redistributed in a dynamic pattern. This constant exchange is an intrinsic property of particles, so that even 'real' particles are nothing more than a little knot of energy which briefly emerges and disappears back into the underlying field. According to quantum field theory, the individual entity is transient and insubstantial. Particles cannot be separated from the empty space around them. Einstein himself recognized that matter itself was 'extremely intense' – a disturbance, in a sense, of perfect randomness – and that the only fundamental reality was the underlying entity – the field itself.[10]

Fluctuations in the atomic world amount to a ceaseless passing back and forth of energy like a ball in a game of pingpong. This energy exchange is analogous to loaning someone a penny: you are a penny poorer, he is a penny richer, until he returns the penny and the roles reverse. This sort of emission and reabsorption of virtual particles occurs not only among photons and electrons, but with all the quantum particles in the universe. The Zero Point Field is a repository of all fields and all ground energy states and all virtual particles – a field of fields. Every exchange of every virtual particle radiates energy. The zero-point energy in any one particular transaction in an electromagnetic field is unimaginably tiny – half a photon's worth.

But if you add up all the particles of all varieties in the universe constantly popping in and out of being, you come up with a vast, inexhaustible energy source – equal to or greater than the energy density in an atomic nucleus – all sitting there unobtrusively in the background of the empty space around us, like one all-pervasive, supercharged backdrop. It has been calculated that the total energy of the Zero Point Field exceeds all energy in matter by a factor of 10^{40}, or 1 followed by 40 zeros.[11] As the great physicist Richard Feynman once described, in attempting to give some idea of this magnitude, the energy in a single cubic meter of space is enough to boil all the oceans of the world.[12]

The Zero Point Field represented two tantalizing possibilities to Hal. Of course, it represented the Holy Grail of energy research. If you could somehow tap into this field, you might have all the energy you would ever need, not simply for fuel on earth, but for space propulsion to distant stars. At the moment, travelling to the nearest star outside our solar system would require a rocket as large as the sun to carry the necessary fuel.

But there was also a larger implication of a vast underlying sea of energy. The existence of the Zero Point Field implied that all matter in the universe was interconnected by waves, which are spread out through time and space and can carry on to infinity, tying one part of the universe to every other part. The idea of The Field might just offer a scientific explanation for many metaphysical notions, such as the Chinese belief in the life force, or *qi*, described in ancient texts as something akin to an energy field. It even echoed the Old Testament's account of God's first dictum: 'Let there be light', out of which matter was created.[13]

Hal was eventually to demonstrate in a paper published by *Physical Review*, one of world's most prestigious physics journals, that the stable state of matter depends for its very existence on this dynamic interchange of subatomic particles with the sustaining zero-point energy field.[14] In quantum theory, a constant problem wrestled with by physicists concerns the issue of why atoms are stable. Invariably, this question would be examined in the laboratory or mathematically tackled using the hydrogen atom. With one electron and one proton, hydrogen is the simplest atom in the universe to dissect. Quantum scientists struggled with the question of why an electron orbits around a proton, like a planet orbiting around the sun. In the solar system, gravity accounts for the stable orbit. But in the atomic world, any moving electron, which carries a charge, wouldn't be stable like an orbiting planet, but would eventually radiate away, or exhaust, its energy and then spiral into the nucleus, causing the entire atomic structure of the object to collapse.

Danish physicist Niels Bohr, another of the founding fathers of quantum theory, sorted the problem by declaring that he wouldn't allow it.[15] Bohr's explanation was that an electron radiates only when it jumps from one orbit to another and that orbits have to have the proper difference in energy to account for any emission of photon light. Bohr made up his own law, which said, in effect, 'there is no energy, it is forbidden. I forbid the electron to collapse'. This dictum and its assumptions led to further assumptions about matter and energy having both wave- and particle-like characteristics, which kept electrons in their place and in particular orbits, and ultimately to the development of quantum mechanics. Mathematically at least, there is no doubt that Bohr was correct in predicting this difference in energy levels.[16]

But what Timothy Boyer had done, and what Hal then perfected, was to show that if you take into account the Zero Point Field, you don't have to rely on Bohr's dictum. You can show mathematically that electrons lose and gain energy constantly from the Zero Point Field in a dynamic equilibrium, balanced at exactly the right orbit. Electrons get their energy to keep going without slowing down because they are refuelling by tapping into these fluctuations of empty space. In other words, the Zero Point Field accounts for the stability of the hydrogen atom – and, by inference, the stability of all matter. Pull the plug on zero-point energy, Hal demonstrated, and all atomic structure would collapse.[17]

Hal also showed by physics calculations that fluctuations of the Zero Point Field waves drive the motion of subatomic particles and that all the motion of all the particles of the universe in turn generates the Zero Point Field, a sort of self-generating feedback loop across the cosmos.[18] In Hal's mind, it was not unlike a cat chasing its own tail.[19] As he wrote in one paper,

> **the ZPF interaction constitutes an underlying, stable 'bottom rung' vacuum state in which further ZPF interaction simply reproduces the existing state on a dynamic-equilibrium basis.[20]**

What this implies, says Hal, is a 'kind of self-regenerating grand ground state of the universe',[21] which constantly refreshes itself and remains a constant unless disturbed in some way. It also means that we and all the matter of the universe are literally connected to the furthest reaches of the cosmos through the Zero Point Field waves of the grandest dimensions.[22]

Much like the undulations of the sea or ripples on a pond, the waves on the subatomic level are represented by periodic oscillations moving through a medium – in this instance the Zero Point Field. They are represented by a classic sideways S, or sine curve, like a jump rope being held at both ends and wiggled up and down. The amplitude of the wave is half the height of the curve from peak to trough, and a single wavelength, or cycle, is one complete oscillation, or the distance between, say, two adjacent peaks or two adjacent troughs. The frequency is the number of cycles in one second, usually measured in hertz, where 1 hertz equals one cycle per second. In the US, our electricity is delivered at a frequency of 60 hertz or cycles per second; in the UK, it is 50 hertz. Cell phones operate on 900 or 1800 megahertz.

When physicists use the term 'phase', they mean the point the wave is at on its oscillating journey. Two waves are said to be in phase when they are both, in effect, peaking or troughing at the same time, even if they have different frequencies or amplitudes. Getting 'in phase' is getting in synch.

One of the most important aspects of waves is that they are encoders and carriers of information. When two waves are in phase, and overlap each other – technically called 'interference' – the combined amplitude of the waves is greater than each individual amplitude. The signal gets stronger. This amounts to an imprinting or exchange of information, called 'constructive interference'. If one is peaking when the other is troughing, they tend to cancel each other out – a process called 'destructive interference'. Once they've collided, each wave contains information, in the form of energy coding, about the other, including all the other information it contains. Interference patterns amount to a constant accumulation of information, and waves have a virtually infinite capacity for storage.

If all subatomic matter in the world is interacting constantly with this ambient ground-state energy field, the subatomic waves of The Field are constantly imprinting a record of the shape of everything. As the harbinger and imprinter of all wavelengths and all frequencies, the Zero Point Field is a kind of shadow of the universe for all time, a mirror image and record of everything that ever was. In a sense, the vacuum is the beginning and the end of everything in the universe.[23]

Although all matter is surrounded with zero-point energy, which bombards a given object uniformly, there have been some instances where disturbances in the field could actually be measured. One such disturbance caused by the Zero Point Field is the Lamb shift, named after American physicist Willis Lamb and developed during the 1940s using wartime radar, which shows that zero-point fluctuations cause electrons to move a bit in their orbits, leading to shifts in frequency of about 1000 megahertz.[24]

Another instance was discovered in the 1940s, when a Dutch physicist named Hendrik Casimir demonstrated that two metal plates placed close together will actually form an attraction that appears to pull them closer together. This is because when two plates are placed near each other, the zero-point waves between the plates are restricted to those that essentially span the gap. Since some wavelengths of the field are excluded, this leads to a disturbance in the equilibrium of the field and the result is an imbalance of energy, with less energy in the gap between the plates than in the outside empty space. This greater energy density pushes the two metal plates together.

Another classic demonstration of the existence of the Zero Point Field is the van der Waals effect, also named after its discoverer, Dutch physicist Johannes Diderik van der Waals. He discovered that forces of attraction and repulsion operate between atoms and molecules because of the way that electrical charge is distributed and, eventually, it was found that this again has to do with a local imbalance in the equilibrium of The Field. This property allows certain gases to turn into liquids. Spontaneous emission, when atoms decay and emit radiation for no known reason, has also been shown to be a Zero Point Field effect.

Timothy Boyer, the physicist whose paper sparked Puthoff in the first place, showed that many of the Through-the-Looking-Glass properties of subatomic matter wrestled with by physicists and leading to the formulation of a set of strange quantum rules could be easily accounted for in classical physics, so long as you also factor in the Zero Point Field. Uncertainty, wave-particle duality, the fluctuating motion of particles: all had to do with the interaction of matter and the Zero Point Field. Hal even began to wonder whether it could account for what remains that most mysterious and vexatious of forces: gravity.

Gravity is the Waterloo of physics. Attempting to work out the basis for this fundamental property of matter and the universe has bedeviled the greatest geniuses of physics. Even Einstein, who was able to describe gravity extremely well through his theory of relativity, couldn't actually explain where it came from. Over the years, many physicists, including Einstein, have tried to assign it an electromagnetic nature, to define it as a nuclear force, or even to give it its own set of quantum rules – all without success. Then, in 1968, the noted Soviet physicist Andrei Sakharov turned the usual assumption on its head. What if gravity weren't an interaction between objects, but just a residual effect? More to the point, what if gravity were an after-effect of the Zero Point Field, caused by alterations in the field due to the presence of matter?[25]

All matter at the level of quarks and electrons jiggles because of its interaction with the Zero Point Field. One of the rules of electrodynamics is that a fluctuating charged particle will emit an electromagnetic radiation field. This means that besides the primary Zero Point Field itself, a sea of these secondary fields exists. Between two particles, these secondary fields cause an attractive source, which Sakharov believed had something to do with gravity.[26]

Hal began pondering this notion. If this were true, where physicists were going wrong was in attempting to establish gravity as an entity in its own right. Instead, it should be seen as a sort of pressure. He began to think of gravity as a kind of long-range Casimir effect, with two objects which blocked some of the waves of the Zero Point Field becoming attracted to each other,[27] or perhaps it was even a long-range van der Waals force, like the attraction of two atoms at certain distances.[28] A particle in the Zero Point Field begins jiggling due to its interaction with the Zero Point Field; two particles not only have their own jiggle, but also get influenced by the field generated by other particles, all doing their own jiggling. Therefore, the fields generated by these particles – which represent a partial shielding of the all-pervasive ground state Zero Point Field – cause the attraction that we think of as gravity.

Sakharov only developed these ideas as a hypothesis; Puthoff went further and began working them out mathematically. He demonstrated that gravitational effects were entirely consistent with zero-point particle motion, what the Germans had dubbed '*zitterbewegung*' or 'trembling motion'.[29] Tying gravity in with zero-point energy solved a number of conundrums that had confounded physicists for many centuries. It answered, for instance, the question of why gravity is weak and why it can't be shielded (the Zero Point Field, which is ever-present, can't be completely shielded itself). It also explained why we can have positive mass and not negative mass. Finally, it brought gravity together with the other forces of physics, such as nuclear energy and electromagnetism, into one cogent unified theory – something physicists had always been eager to do but had always singularly failed at.

Hal published his theory of gravity to polite and restrained applause. Although no one was rushing to duplicate his data, at least he wasn't being ridiculed, even though what he'd been saying in these papers in essence unsettled the entire bedrock of twentieth-century physics. Quantum physics most famously claims that a particle can also simultaneously be a wave unless observed and then measured, when all its tentative possibilities collapse into a set entity. With Hal's theory, a particle is always a particle but its state just seems indeterminate because it is constantly interacting with this background energy field. Another quality of subatomic particles such as electrons taken as a given in quantum theory is 'nonlocality' – Einstein's 'spooky action at a distance'. This quality may also be accounted for by the Zero Point Field. To Hal, it was analogous to two sticks planted in the sand at the edge of the ocean about to be hit by a rolling wave. If you didn't know about the wave, and both sticks fell down because of it one after the other, you might think one stick had affected the other at a distance and call that a non-local effect. But what if it were zero-point fluctuation that was the underlying mechanism acting on quantum entities and causing one entity to affect the other?[30] If that were true, it meant every part of the universe could be in touch with every other part instantaneously.

While continuing with other work at SRI, Hal set up a small lab in Pescadero, in the foothills of the northern California coastline, within the home of Ken Shoulders, a brilliant lab engineer he'd known from years before whom he'd lately recruited to help him. Hal and Ken began working on condensed charge technology, a sophisticated version of scuffling your foot across a carpet and then getting a shock when you touch metal. Ordinarily, electrons repel each other and don't like to be pushed too closely together. However, you can tightly cluster electronic charge if you calculate in the Zero Point Field, which at some point will begin to push electrons together like a tiny Casimir force. This enables you to develop electronics applications in very tiny spaces.

Hal and Ken began coming up with gadget applications that would use this energy and then patenting their discoveries. Eventually they would invent a special device that could fit an X-ray device at the end of a hypodermic needle, enabling medics to take pictures of body parts in tiny crevices, and then a high-frequency signal generator radar device that would allow radar to be generated from a source no larger than a plastic credit card. They would also be among the first to design a flat-panel television, the width of a hanging picture. All their patents were accepted with the explanation that the ultimate source of energy 'appears to be the zero-point radiation of the vacuum continuum'.[31]

Hal and Ken's discoveries were given an unexpected boost when the Pentagon, which rates new technologies in order of importance to the nation, listed condensed-charge technology, as zero-point energy research was then termed, as number 3 on the National Critical Issue List, only after stealth bombers and optical computing. A year later, condensed-charge technology would move into the number two slot. The Interagency Technological Assessment Group was convinced that Hal was onto something important to the national interest and that aerospace could develop further only if energy could be extracted from the vacuum.

With the US government endorsing their work, Puthoff and Shoulders could have had their pick of private companies willing to fund their research. Eventually, in 1989, they went with Boeing, which was interested in their tiny radar device and planned to fund its development on the back of a large project. The project languished for a couple of years, and then Boeing lost the funding. Most of the other companies demanded a full-scale prototype before they would fund the project. Hal decided to set up his own company to develop the X-ray device. He got halfway along that route before it occurred to him that he was about to take an unwelcome detour. It might make him a lot of money, but he was only interested in the project for the money he could use to fund his energy research. Setting up and running this company would take at least 10 years out of his life, he figured, much as Bill's family business had consumed a decade of his. Far better, he thought, simply to look for funding for the energy research itself. Hal made the decision then and there. He would keep his eye firmly on the altruistic goal he'd started with – and would eventually bet his entire career on it. First service, then glory and last, if at all, remuneration.

Hal would wait nearly 20 years for anyone else to replicate and expand his theories. His confirmation came with a telephone message, left at 3 a.m., that would seem braggardly, ridiculous even, to most physicists. Bernie Haisch had been wrapping up a few last details in his Lockheed office in Palo Alto, getting ready to embark on a research fellowship he'd got at the Max Planck Institute at Garching, Germany. An astrophysicist at Lockheed, Bernie was looking forward to spending the rest of his summer doing research on the X-ray emission of stars and considered himself lucky to have landed the opportunity. Bernie was an odd hybrid, a formal and cautious manner belying a private expressiveness which found its outlet in writing folk songs. But in the laboratory he was as little given to hyperbole as his friend Alfonso Rueda, a noted physicist and applied mathematician at the California State University in Long Beach, who'd left the message.

Physicists were hardly noted for a sense of humor about their work, and the Colombian was a quiet detail man, certainly not given to boastfulness. Maybe it was Rueda's idea of a practical joke.

The message left on Haisch's answering machine had said, 'Oh my God, I think I've just derived $F = ma$.'

To a physicist, this announcement was analogous to claiming to have worked out a mathematical equation to prove God. In this case, God was Newton and $F = ma$ the First Commandment. $F = ma$ was a central tenet in physics, postulated by Newton in his *Principia*, the Holy Bible of classical physics, in 1687, as the fundamental equation of motion. It was so central to physical theory that it was a given, a postulate, not something provable, but simply assumed to be true, and never argued with. Force equals mass (or inertia) times acceleration. Or, the acceleration you get is inversely proportional to mass for any given force. Inertia – the tendency of objects to stay put and be hard to get moving, and then once moving, hard to stop – fights your ability to increase the speed of an object. The bigger the object, the more force is needed to get it moving. The amount of effort it takes to send a flea flying across a tennis court will not begin to shift a hippopotamus.

The point was, no one mathematically *proved* a commandment. You use it to build an entire religion upon. Every physicist since Newton took that to be a fundamental assumption and built theory and experiment based upon this bedrock. Newton's postulate essentially had defined inertial mass and laid the foundation of physical mechanics for the last 300 years. We all know it to be true, even though nobody could actually prove it.[32]

And now Alfonso Rueda was claiming, in his phone message, that this very equation, the most famous in all of physics besides $E = mc^2$, was the end result of a fevered mathematical calculation that he had been grinding away at late into the night for many months. He would mail details to Bernie in Germany.

Although he was embroiled in his aerospace work, Bernie had read some of Hal Puthoff's papers and himself got interested in the Zero Point Field, largely as a source of energy for distant space travel. Bernie had been inspired by the work of British physicist Paul Davies and William Unruh of the University of British Columbia. The pair had found that if you move at a constant speed through the vacuum, it all looks the same. But as soon as you start to accelerate, the vacuum begins to appear like a lukewarm sea of heat radiation from your perspective as you move.

Bernie began wondering if inertia – like this heat radiation – is caused by acceleration through the vacuum.[33]

Then, at a conference, he'd met Rueda, a well-known physicist with an extensive background in high-level mathematics, and after much encouragement and prodding from Bernie, the ordinarily dour Rueda began to work through the analysis involving the Zero Point Field and an idealized oscillator, a fundamental device used to work through many classic problems in physics. Although Bernie had his own technical expertise, he needed a high-level mathematician to do the calculations. He'd been intrigued by Hal's work on gravity and considered that there might be a connection between inertia and the Zero Point Field.

After many months, Rueda had finished the calculations. What he found was that an oscillator forced to accelerate through the Zero Point Field will experience resistance, and that this resistance will be proportional to acceleration. It looked, for all the world, as though they'd just been able to show why $F = ma$. No longer was it simply because Newton had deigned to define it as such. If Alfonso was right, one of the fundamental axioms of the world had been reduced to something you could derive from electrodynamics. You didn't have to assume anything. You could prove that Newton was right simply by taking account of the Zero Point Field.

Once Bernie had received Rueda's calculations, he contacted Hal Puthoff, and the three of them decided to work together. Bernie wrote it up as a very long paper. After some foot-dragging, *Physical Review*, a very prestigious mainstream physics journal, published the paper unchanged in February 1994.[34] The paper demonstrated that the property of inertia possessed by all objects in the physical universe was simply resistance to being accelerated through the Zero Point Field. In their paper they showed that inertia is what is termed a Lorentz force – a force that slows particles moving through a magnetic field. In this instance, the magnetic field is a component of the Zero Point Field, reacting with the charged subatomic particles. The larger the object, the more particles it contains and the more it is held stationary by the field.

What this was basically saying is that the corporeal stuff we call matter and to which all physicists since Newton have attributed an innate mass was an illusion. All that was happening was that this background sea of energy was opposing acceleration by gripping on to the subatomic particles whenever you pushed on an object. Mass, in their eyes, was a 'bookkeeping' device, a 'temporary place holder' for a more general quantum vacuum reaction effect.[35]

Hal and Bernie also realized that their discovery had a bearing on Einstein's famous equation $E = mc^2$. The equation has always implied that energy (one distinct physical entity in the universe) turns into mass (another distinct physical entity). They now saw that the relationship of mass to energy was more a statement about the energy of quarks and electrons in what we call matter caused by interaction with the Zero Point Field fluctuations. What they were all getting at, in the mild-mannered, neutral language of physics, was that matter is not a fundamental property of physics. The Einstein equation was simply a recipe for the amount of energy necessary to create the appearance of mass. It means that there aren't two fundamental physical entities – something material and another immaterial – but only one: energy. Everything in your world, anything you hold in your hand, no matter how dense, how heavy, how large, on its most fundamental level boils down to a collection of electric charges interacting with a background sea of electromagnetic and other energetic fields – a kind of electromagnetic drag force. As they would write later, mass was not equivalent to energy; mass *was* energy.[36] Or, even more fundamentally, there is no mass. There is only charge.

Noted science writer Arthur C. Clarke later predicted that the Haisch–Rueda–Puthoff paper would one day be regarded as a 'landmark'[37], and in *3001: The Final Odyssey*, gave a nod to their contribution by creating a spacecraft powered by an inertia-cancelling drive known as the SHARP drive (an acronym for 'Sakharov, Haisch, Alfonso Rueda and Puthoff').[38] As Clarke wrote, in justifying his immortalization of their theory:

> It addresses a problem so fundamental that it is normally taken for granted, with a that's-just-the-way-the-universe-is-made shrug of the shoulders.
>
> The question HR & P asked is: 'What gives an object mass (or inertia) so that it requires an effort to start it moving, and exactly the same effort to restore it to its original state?
>
> Their provisional answer depends on the astonishing and – outside the physicists' ivory towers – little-known fact that so-called empty space is actually a cauldron of seething energies – the Zero Point Field . . . HR & P suggest that both inertia and gravitation are electromagnetic phenomena resulting from interaction with this field.

> There have been countless attempts, going all the way back to Faraday, to link gravity and magnetism, and although many experimenters have claimed success, none of their results has ever been verified. However, if HR & P's theory can be proved, it opens up the prospect – however remote – of anti-gravity 'space drives' and the even more fantastic possibility of controlling inertia. This could lead to some interesting situations: if you gave someone the gentlest touch, they would promptly disappear at thousands of kilometres an hour, until they bounced off the other side of the room a fraction of a millisecond later. The good news is that traffic accidents would be virtually impossible: automobiles – and passengers – could collide harmlessly at any speed.[39]

Elsewhere, in an article about future space travel, Clarke wrote: 'If I was a NASA administrator . . . I'd get my best, brightest and youngest (no one over 25 need apply) to take a long, hard look at Puthoff *et al.*'s equations.'[40] Later, Haisch, Rueda and Daniel Cole of IBM would publish a paper showing that the universe owes its very structure to the Zero Point Field. In their view, the vacuum causes particles to accelerate, which in turn causes them to agglutinate into concentrated energy, or what we call matter.[41]

In a sense, the SHARP team had done what Einstein himself had not done.[42] They had proved one of the most fundamental laws of the universe, and found an explanation for one of its greatest mysteries. The Zero Point Field had been established as the basis of a number of fundamental physical phenomena. Bernie Haisch, with his NASA background, had his sights firmly on the possibilities open to space travel of having inertia, mass and gravity all tied to this background sea of energy. Both he and Hal received funding to develop an energy source extracted from the vacuum, in Bernie's case from a NASA eager to advance space travel.

If you could extract energy from the Zero Point Field wherever you are in the universe, you wouldn't have to carry fuel with you, but could just set sail in space and tap into the Zero Point Field – a kind of universal wind – whenever you needed to. Hal Puthoff had showed in another paper, also with Daniel Cole from IBM, that in principle there was nothing in the laws of thermodynamics to exclude the possibility of extracting energy from it.[43] The other idea was to manipulate the waves of the Zero Point Field, so that they would act like a unilateral force, pushing your vehicle along. Bernie imagined that at some point in the future, you might be able to just set your zero-point transducer (wave transformer) and go. But perhaps even more exotic, if you could modify or turn off inertia you might be able to set off a rocket with very low energy, but just modify the forces that stop it from moving. Or use a very fast rocket, but modify the inertia of the astronauts so that they wouldn't be flattened by G forces. And if you could somehow turn off gravity, you could change the weight of the rocket or the force required to accelerate it.[44] The possibilities were endless.

But that wasn't the only aspect of zero-point energy with potential. In some of his other work, Hal had come across studies of levitation. The modern cynical view was that these feats were performed by sleight of hand, or were the hallucinations of religious fanatics. Nevertheless, many of the people who'd attempted to debunk these feats had failed. Hal found exquisite notes about the events. To the physicist in him, who always needed to take a given situation apart and examine the pieces, as he had in his youth with ham radios, what was being described appeared to be a relativistic phenomenon. Levitation is categorized as psychokinesis, the ability of humans to make objects (or themselves) move in the absence of any known force. The recorded instances of levitation that Hal had stumbled across only seemed possible in a physics sense if gravity had somehow been manipulated. If these vacuum fluctuations, considered so meaningless by most quantum physicists, did amount to something that could be harnessed at will, whether for automobile fuel or to move objects just by focusing one's attention on them, then the implications not only for fuel but for every aspect of our lives were enormous. It might be the closest we have to what in *Star Wars* was called 'The Force'.

In his professional work, Hal was careful to stay firmly within the confines of conservative physics theory. Nevertheless, privately he was beginning to understand the metaphysical implications of a background sea of energy. If matter wasn't stable, but an essential element in an underlying ambient, random sea of energy, he thought, then it should be possible to use this as a blank matrix on which coherent patterns could be written, particularly as the Zero Point Field had imprinted everything that ever happened in the world through wave interference encoding. This kind of information might account for coherent particle and field structures. But there might also be an ascending ladder of other possible information structures, perhaps coherent fields around living organisms, or maybe this acts as a non-biochemical 'memory' in the universe. It might even be possible to organize these fluctuations somehow through an act of will.[45] As Clarke had written, 'We may already be tapping this in a very small way: it may account for some of the anomalous 'over-unity' results now being reported from many experimental devices, by apparently reputable engineers.'[46]

Hal, like Bernie, was first and last a physicist who didn't let his mind run away with itself, but when he did allow himself a few moments of speculation, he realized that this represented nothing less than a unifying concept of the universe, which showed that everything was in some sort of connection and balance with the rest of the cosmos. The universe's very currency might be learned information, as imprinted upon this fluid, mutable field of information. The Field demonstrated that the real currency of the universe – the very reason for its stability – is an *exchange* of energy. If we were all connected through The Field, then it just might be possible to tap into this vast reservoir of energy information and extract information from it. With such a vast energy bank to be harnessed, virtually anything was possible – that is, if human beings had some sort of quantum structure allowing them access to it. But there was the stumbling block. That would require that our bodies operated according to the laws of the quantum world.

CHAPTER THREE

Beings of Light

FRITZ-ALBERT POPP THOUGHT he had discovered a cure for cancer. It was 1970, a year before Edgar Mitchell had flown to the moon, and Popp, a theoretical biophysicist at the University of Marburg in Germany, had been teaching radiology, the interaction of electromagnetic radiation on biological systems. He'd been examining benzo[a]pyrene, a polycyclic hydrocarbon known to be one of the most lethal carcinogens to humans and had illuminated it with ultraviolet light.

Popp played around with light a lot. He'd been fascinated by the effect of electromagnetic radiation on living systems ever since he'd been a student at the University of Würzburg. During his time as an undergraduate he'd studied in the house, sometimes even in the very room, where Wilhelm Röntgen had accidentally stumbled on the fact that rays of a certain frequency could produce pictures of the hard structures of the body.

Popp had been trying to determine what effect you'd get if you excited this deadly compound with ultraviolet (UV) light. What he discovered was that benzo[a]pyrene had a crazy optical property. It absorbed the light but then re-emitted it at a completely different frequency, like some CIA operative intercepting a communication signal from the enemy and jumbling it up. This was a chemical which doubled as a biological frequency scrambler. Popp then performed the same test on benzo[e]pyrene, another polycyclic hydrocarbon, which is virtually identical in every way to benzo[a]pyrene save for a tiny alteration in its molecular makeup. This tiny difference in one of the compound rings was critical as it rendered benzo[e]pyrene harmless to humans. With this particular chemical, the light passed right through the substance unaltered.

Popp kept puzzling over this difference and kept playing around with light and compounds. He performed his test on thirty-seven other chemicals, some cancer-causing, some not. After a while, it got so that he could predict which substances could cause cancer. In every instance, the compounds that were carcinogenic took the UV light, absorbed it, and changed the frequency.

There was another odd property of these compounds. Each of the carcinogens reacted only to the light at a specific wavelength – 380 nanometres. Popp kept wondering why a cancer-causing substance would be a light scrambler. He began reading the scientific literature, specifically about human biological reactions, and came across information about a phenomenon called 'photo-repair'. It is very well known from biological laboratory experiments that if you can blast a cell with UV light so that 99 per cent of the cell, including its DNA, is destroyed, you can almost entirely repair the damage in a single day just by illuminating the cell with the same wavelength of a very weak intensity. To this day, conventional scientists don't understand this phenomenon, but nobody has disputed it. Popp also knew that patients with a skin condition called xeroderma pigmentosum eventually die of skin cancer because their photo-repair system doesn't work and so doesn't repair solar damage. Popp was shocked to learn that photo-repair works most efficiently at 380 nanometres – the very same wavelength the cancer-causing compounds would react to and scramble.

This was where Popp made his logical leap. Nature was too perfect for this to be simple coincidence. If the carcinogens only react to this wavelength, it must somehow be linked to photo-repair. If so, this would mean that there must be some light in the body responsible for photo-repair. A cancerous compound must cause cancer because it permanently blocks this light and scrambles it, so photo-repair can't work anymore.

Popp was profoundly taken aback by the thought of it all. He decided there and then that this was where his future work would lie. He wrote the paper up, but told few people about it, and was pleased, but not really surprised, when a prestigious journal on cancer agreed to publish it.[1] In the months before his paper was published, Popp was highly impatient, worried that his idea would be stolen. Any careless disclosure of his to the casual observer might send the listener off to patent Popp's discovery. As soon as the scientific community realized he had discovered a cure for cancer, he would be one of the most celebrated scientists of his day. It was his first foray into a new area of science, and it was going to land him the Nobel prize.

Popp, after all, was used to accolades. Up until that point he'd won nearly every prize you could be awarded in academic life. He'd even picked up the Röntgen prize for his undergraduate diploma work, which consisted of building a small particle accelerator. This prize, named after Popp's hero, Wilhelm Röntgen, is given each year to the top undergraduate in physics at the University of Würzburg. Popp had studied like a young man possessed. He'd finished his examinations far earlier than the other students. He was awarded his PhD in theoretical physics in record time. The postgraduate work required for German professorships, a five-year proposition for most academics, took Popp just a little more than two years. At the time of his discovery, Popp was already celebrated among his peers for being a whiz kid, not only because of his ability but also because of his dashing, youthful looks.

When his paper was published, Popp was 33 and good-looking, with the set jaw and direct steel-blue gaze of a Hollywood swashbuckler and a boyish face always assumed to be years younger. Even his wife, who was seven years younger than him, was often mistaken as the senior partner. And indeed, there was something of the swashbuckler about him; he had a reputation among his fellow students as the best fencer on campus – a reputation which had been tested in various duels, one of which had left him with a gash all along the left side of his head.

Popp's looks and manner belied his seriousness of purpose. Like Edgar Mitchell, he was a philosopher as much as a scientist. Even as a tiny child he'd been trying to make sense of the world, to find some general solution he could apply to everything in his life. He'd even planned to study philosophy until a teacher persuaded him that physics might be a more fertile territory if he required some single equation that held the key to life. Nevertheless, classical physics, with its assertion of reality as a phenomenon independent of the observer, had left him profoundly suspicious. Popp had read Kant and believed, like the philosopher, that reality was the creation of living systems. The observer must be central to the creation of his world.

Popp was celebrated for his paper. The Deutsche Krebsforschungszentrum (German Cancer Research Center) in Heidelberg invited him to speak before fifteen of the world's leading cancer specialists during an eight-day conference on all aspects of cancer. The invitation to speak among such exclusive company was an incredible opportunity, and it increased his prestige on his university campus. He arrived in a brand new suit, the most elegant presence at the colloquium, but he was the poorest speaker, struggling with his English to make his voice heard.

In his presentation as well as his paper, Popp's science was unassailable, save for one detail: it assumed that a weak light of 380 nanometres was somehow being produced in the body. To the cancer researchers, this one detail was some kind of a joke. Don't you think if there were light in the body, they told him, somebody, somewhere would have noticed it by now?

Only a single researcher, a photochemist from the Madame Curie Institute, working on the carcinogenic activity of molecules, was convinced that Popp was right. She invited Popp to work with her in Paris, but would herself die of cancer before he could join her.

The cancer researchers challenged Popp to come up with evidence, and he was ready with a counter challenge. If they would help him build the right equipment, then he would show them where the light was coming from.

Not long after, Popp was approached by a student named Bernhard Ruth, who asked Popp to supervise his work for his PhD dissertation.

'Sure,' said Popp, 'if you can show that there is light in the body.'

Ruth thought it a ridiculous suggestion. Of course, there isn't light in the body.

'Okay,' said Popp. 'So show me evidence that there isn't light, and you can get your PhD.'

This meeting was fortuitous for Popp because Ruth happened to be an excellent experimental physicist. He set to work building equipment which would demonstrate, once and for all, that no light was emanating from the body. Within two years he'd produced a machine resembling a big X-ray detector (EMI 9558QA selected typed), which employed a photomultiplier, enabling it to count light, photon by photon. To this day it is still one of the best pieces of equipment in the field. The machine had to be highly sensitive because it would be measuring what Popp assumed would be extremely weak emissions.

In 1976, they were ready for their first test. They'd grown cucumber seedlings, which are among the easiest of plants to cultivate, and put them in the machine. The photomultiplier picked up that photons, or light waves, of a surprisingly high intensity were being emitted from the seedlings. Ruth was highly sceptical. This had something to do with chlorophyll, he argued – a position Popp shared. They decided that with their next test – some potatoes – they would grow the seedling plants in the dark, so they could not undergo photosynthesis. Nevertheless, when placed in the photomultiplier, these potatoes registered an even higher intensity of light.[2] It was impossible that the effect had anything to do with photosynthesis, Popp realized. What's more, these photons in the living systems he'd examined were more coherent than anything he'd ever seen.

In quantum physics, quantum coherence means that subatomic particles are able to cooperate. These subatomic waves or particles not only know about each other, but also are highly interlinked by bands of common electromagnetic fields, so that they can communicate together. They are like a multitude of tuning forks that all begin resonating together. As the waves get into phase or synch, they begin acting like one giant wave and one giant subatomic particle. It becomes difficult to tell them apart. Many of the weird quantum effects seen in a single wave apply to the whole. Something done to one of them will affect the others.

Coherence establishes communication. It's like a subatomic telephone network. The better the coherence, the finer the telephone network and the more refined wave patterns have a telephone. The end result is also a bit like a large orchestra. All the photons are playing together but as individual instruments that are able to carry on playing individual parts. Nevertheless, when you are listening, it's difficult to pick out any one instrument.

What was even more amazing was that Popp was witnessing the highest level of quantum order, or coherence, possible in a living system. Usually, this coherence – called a Bose–Einstein condensate – is only observed in material substances such as superfluids or superconductors studied in the laboratory in very cold places – just a few degrees above absolute zero – and not in the hot and messy environment of a living thing.

Popp began thinking about light in nature. Light, of course, was present in plants, the source of energy used during photosynthesis. When we eat plant foods, it must be, he thought, that we take up the photons and store them. Say that we consume some broccoli. When we digest it, it is metabolized into carbon dioxide (CO_2) and water, plus the light stored from the sun and present in photosynthesis. We extract the CO_2 and eliminate the water, but the light, an electromagnetic wave, must get stored. When taken in by the body, the energy of these photons dissipates so that it is eventually distributed over the entire spectrum of electromagnetic frequencies, from the lowest to the highest. This energy becomes the driving force for all the molecules in our body.

Photons switch on the body's processes like a conductor launching each individual instrument into the collective sound. At different frequencies they perform different functions. Popp found with experimentation that molecules in the cells would respond to certain frequencies and that a range of vibrations from the photons would cause a variety of frequencies in other molecules of the body. Light waves also answered the question of how the body could manage complicated feats with different body parts instantaneously or do two or more things at once. These 'biophoton emissions', as he was beginning to call them, could provide a perfect communication system, to transfer information to many cells across the organism. But the single most important question remained: where were they coming from?

A particularly gifted student of his talked him into trying an experiment. It is known that when you apply a chemical called ethidium bromide to samples of DNA, the chemical squeezes itself into the middle of the base pairs of the double helix and causes it to unwind. The student suggested that, after applying the chemical, he and Popp try measuring the light coming off the sample. Popp discovered that the more he increased the concentration of the chemical, the more the DNA unwound, but also the stronger the intensity of light. The less he put in, the lower the light emission.[3] He also found that DNA was capable of sending out a large range of frequencies and that some frequencies seemed linked to certain functions. If DNA were storing this light, it would naturally emit more light once it was unwound.

These and other studies demonstrated to Popp that one of the most essential stores of light and sources of biophoton emissions was DNA. DNA must be like the master tuning fork in the body. It would strike a particular frequency and certain other molecules would follow. It was altogether possible, he realized, that he might have stumbled upon the missing link in current DNA theory that could account for perhaps the greatest miracle of all in human biology: the means by which a single cell turns into a fully formed human being.

One of the greatest mysteries of biology is how we and every other living thing take geometric shape. Modern scientists mostly understand how we have blue eyes or grow to six foot one, and even how cells divide. What is far more elusive is the manner by which these cells know exactly where to place themselves in each stage of the building process, so that an arm becomes an arm rather than a leg, as well as the very mechanism which gets these cells to organize and assemble themselves together into something resembling a three-dimensional human form.

The usual scientific explanation has to do with the chemical interactions between molecules and with DNA, the coiled double helix of genetic coding that holds a blueprint of the body's protein and amino acids. Each DNA helix or chromosome – and the identical twenty-six pairs exist in every one of the thousand million million cells in your body[4] – contains a long chain of nucleotides, or bases, of four different components (shortened to ATCG) arranged in a unique order in every human body. The most favored idea is that there exists a genetic 'program' of genes operating collectively to determine shape, or, in the view of neo-Darwinists such as Richard Dawkins, that ruthless genes, like Chicago thugs, have powers to create form and that we are 'survival machines' – robot vehicles blindly programmed to preserve the selfish molecules known as genes.[5]

This theory promotes DNA as the Renaissance man of the human body – architect, master builder and central engine room – whose tool for all this amazing activity is a handful of the chemicals which make proteins. The modern scientific view is that DNA somehow manages to build the body and spearhead all its dynamic activities just by selectively turning off and on certain segments, or genes, whose nucleotides, or genetic instructions, select certain RNA molecules, which in turn select from a large alphabet of amino acids the genetic 'words' which create specific proteins. These proteins supposedly are able to both build the body and to switch on and off all the chemical processes inside the cell which ultimately control the running of the body.

Undoubtedly proteins do play a major role in bodily function. Where the Darwinists fall short is in explaining exactly how DNA knows when to orchestrate this and also how these chemicals, all blindly bumping into each other, can operate more or less simultaneously. Each cell undergoes, on average, some 100,000 chemical reactions per second – a process that repeats itself simultaneously across every cell in the body. At any given second, billions of chemical reactions of one sort or another occur. Timing must be exquisite, for if any one of the individual chemical processes in all the millions of cells in the body is off by a fraction, humans would blow themselves up in a matter of seconds. But what the rank and file among geneticists have not addressed is that if DNA is the control room, what is the feedback mechanism which enables it to synchronize the activities of individual genes and cells to carry out systems in unison? What is the chemical or genetic process that tells certain cells to grow into a hand and not a foot? And which cell processes happen at which time?

If all these genes are working together like some unimaginably big orchestra, who or what is the conductor? And if all these processes are due to simple chemical collision between molecules, how can it work anywhere near rapidly enough to account for the coherent behaviours that live beings exhibit every minute of their lives?

When a fertilized egg starts to multiply and produce daughter cells, each begins adopting a structure and function according to its eventual role in the body. Although every daughter contains the same chromosomes with the same genetic information, certain types of cells immediately 'know' to use different genetic information to behave differently from others and so certain genes must 'know' that it is their turn to be played, rather than the rest of the pack. Furthermore, somehow these genes know how many of each type of cell must be produced in the right place. Each cell, furthermore, needs to be able to know about its neighboring cells to work out how it fits into the overall scheme. This requires nothing less than an ingenious method of communication between cells at a very early stage of the embryo's development and the same sophistication every moment of our lives.

Geneticists appreciate that cell differentiation utterly depends on cells knowing how to differentiate early on and then somehow remembering that they are different and passing on this vital piece of information to subsequent generations of cells. At the moment, scientists shrug their shoulders as to how this might all be accomplished, particularly at such a rapid pace.

Dawkins himself admits: 'Exactly how this eventually leads to the development of a baby is a story which will take decades, perhaps centuries, for embryologists to work out. But it is a fact that it does.'[6]

In other words, like policemen desperate to close a case, scientists have arrested the most likely suspect without bothering with the painstaking process of gathering proof. The details of this absolute certainty, of how proteins might accomplish this all on their own, are left decidedly imprecise.[7] As for the orchestration of cell processes, biochemists never actually ask the question.[8]

British biologist Rupert Sheldrake has mounted one of the most constant and vociferous challenges to this approach, arguing that gene activation and proteins no more explain the development of form than delivering building materials to a building site explains the structure of the house built there. Current genetic theory also doesn't explain, he says, how a developing system can self-regulate, or grow normally in the course of development if a part of the system is added or removed, and doesn't explain how an organism regenerates – replacing missing or damaged structures.[9]

In a rush of fevered inspiration while at an ashram in India, Sheldrake worked out his hypothesis of formative causation, which states that the forms of self-organizing living things – everything from molecules and organisms to societies and even entire galaxies – are shaped by morphic fields. These fields have a morphic resonance – a cumulative memory – of similar systems through cultures and time, so that species of animals and plants 'remember' not only how to look but also how to act. Rupert Sheldrake uses the term 'morphic fields' and an entire vocabulary of his own making to describe the self-organizing properties of biological systems, from molecules to bodies to societies. 'Morphic resonance', is, in his view, 'the influence of like upon like through space and time'. He believes these fields (and he thinks there are many of them) are different from electromagnetic fields because they reverberate across generations with an inherent memory of the correct shape and form.[10] The more we learn, the easier it is for others to follow in our footsteps.

Sheldrake's theory is beautifully and simply worked out. Nevertheless, by his own admission, it doesn't explain the physics of how this might all be possible, or how all these fields might store this information.[11]

In biophoton emissions, Popp believed that he had an answer to the question of morphogenesis as well as '*gestaltbildung*' – cell coordination and communication – which only could occur in a holistic system, with one central orchestrator. Popp showed in his experiments that these weak light emissions were sufficient to orchestrate the body. The emissions had to be of low intensity because these communications were occurring on a quantum level, and higher intensities would be felt only in the world of the large.

When Popp began researching this area, he realized he was standing on the shoulders of many others, whose work suggested a field of electromagnetic radiation which somehow guides the growth of the cellular body. It was the Russian scientist Alexander Gurwitsch who had to be credited with first discovering what he called 'mitogenetic radiation' in onion roots in the 1920s. Gurwitsch postulated that a field, rather than chemicals alone, was probably responsible for the structural formation of the body. Although Gurwitsch's work was largely theoretical, later researchers were able to show that a weak radiation from tissues stimulates cell growth in neighboring tissues of the same organism.[12]

Other early studies of this phenomenon – now repeated by many scientists – were carried out in the 1940s by neuroanatomist Harold S. Burr from Yale University, who studied and measured electrical fields around living things, specifically salamanders. Burr discovered that salamanders possessed an energy field shaped like an adult salamander, and that this blueprint even existed in an unfertilized egg.[13]

Burr also discovered electrical fields around all sorts of organisms, from molds, to salamanders and frogs, to humans,[14] Changes in the electrical charges appeared to correlate with growth, sleep, regeneration, light, water, storms, the development of cancer – even the waxing and waning of the moon.[15] For instance, in his experiments with plant seedlings, he discovered electrical fields which resembled the eventual adult plant.

Another of the early interesting experiments was carried out in the early 1920s by Elmer Lund, a researcher at the University of Texas, on hydras, the tiny aquatic animal possessing up to twelve heads capable of regenerating. Lund (and later others) found that he could control regeneration by applying tiny currents through the hydra's body. By using a current strong enough to override the organism's own electrical force, Lund could cause a head to form where a tail should be. In later studies in the 1950s, G. Marsh and H. W. Beams discovered that if voltages were high enough, even a flatworm would begin reorganizing – *the head would turn into a tail and vice versa*. Yet other studies have demonstrated that very young embryos, shorn of their nervous system, and grafted onto a healthy embryo, will actually survive, like a Siamese twin, on the back of the healthy embryos. Still other experiments have shown that regeneration can even be reversed by passing a small current through a salamander's body.[16]

Orthopaedist Robert O. Becker mainly engaged in work concerning attempts to stimulate or speed up regeneration in humans and animals. However, he has also published many accounts of experiments in the *Journal of Bone and Joint Surgery* demonstrating a 'current of injury' – where animals such as salamanders with amputated limbs develop a change of charge at the site of the stump, whose voltage climbs until the new limb appears.[17]

Many biologists and physicists have advanced the idea that radiation and oscillating waves are responsible for synchronizing cell division and sending chromosomal instructions around the body. Perhaps the best known of these, Herbert Fröhlich, of the University of Liverpool, recipient of the prestigious Max Planck Medal, an annual award of the German Physical Society to honour the career of an outstanding physicist, was one of the first to introduce the idea that some sort of collective vibration was responsible for getting proteins to cooperate with each other and carry out instructions of DNA and cellular proteins. Fröhlich even predicted that certain frequencies (now termed 'Fröhlich frequencies') just beneath the membranes of the cell could be generated by vibrations in these proteins. Wave communication was supposedly the means by which the smaller activities of proteins, the work of amino acids, for instance, would be carried out and a good way to synchronize activities between proteins and the system as a whole.[18]

In his own studies, Fröhlich had shown that once energy reaches a certain threshold, molecules begin to vibrate in unison, until they reach a high level of coherence. The moment molecules reach this state of coherence, they take on certain qualities of quantum mechanics, including nonlocality. They get to the point where they can operate in tandem.[19]

The Italian physicist Renato Nobili of the Universita degli Studi di Padova amassed experimental proof that electromagnetic frequencies occur in animal tissues. In experiments he found that the fluid in cells holds currents and wave patterns and that these correspond with wave patterns picked up by electroencephalogram (EEC) readings in the brain cortex and scalp.[20] Russian Nobel prize winner Albert Szent-Györgyi postulated that protein cells act as semiconductors, preserving and passing along the energy of electrons as information.[21]

However, most of this research, including Gurwitsch's initial work, had largely been ignored, mostly because there was no equipment sensitive enough to measure these tiny particles of light before the invention of Popp's machine. Furthermore, any notions of the use of radiation in cellular communication were utterly swept aside in the middle of the twentieth century, with the discovery of hormones and the birth of biochemistry, which proposed that everything could be explained by hormones or chemical reactions.[22]

By the time that Popp had his light machine, he was more or less on his own with regard to a radiation theory of DNA. Nevertheless, he doggedly pressed on with his experiments, learning more about the properties of this mysterious light. The more he tested, the more he discovered that all living things – from the most basic of plants or animals, to human beings in all their sophisticated complexity – emitted a permanent current of photons, from only a few to hundreds. The number of photons emitted seemed to be linked to an organism's position on the evolutionary scale:

the more complex the organism, the fewer photons being emitted. Rudimentary animals or plants tended to emit 100 photons per square centimetre per second, at a wavelength of 200 to 800 nanometres, corresponding to a very high frequency of electromagnetic wave, well within the visible light range, whereas humans would emit only ten photons in the same area, time and frequency. He also discovered something else curious. When light was shone on living cells, the cells would take this light and after a certain delay, shine intensely – a process called 'delayed luminescence'. It occurred to Popp that this could be a corrective device. The living system had to maintain a delicate equilibrium of light. In this instance, when it was being bombarded with too much light, it would reject the excess.

Very few places in the world can claim to be pitch black. The only appropriate candidates would be an enclosure where only a handful of photons remain. Popp possessed such a place, a room so dark that only the barest few photons of light per minute could be detected in it. This was the only fit laboratory in which to measure the light of human beings. He began studying the patterns of biophoton emissions of some of his students. In one series of studies, he had one of his experimenters – a 27-year-old healthy young woman – sit in the room every day for nine months, while he took photon readings of a small area of her hand and forehead. Popp then analysed the data, and discovered, to his surprise, that the light emissions followed certain set patterns – biological rhythms at 7, 14, 32, 80 and 270 days, when the emissions were identical, even after one year. Emissions for both the left and right hands were also correlated. If there was an increase in the photons coming off the right hand, so there would be a similar increase in the those of the left hand. On a subatomic level, the waves of each hand were in phase. In terms of light, the right hand knew what the left hand was doing.

Emissions also seemed to follow other natural biological rhythms; similarities were noted by day or night, by week, by month, as though the body were following the world's biorhythms as well as its own.

So far, Popp had studied only healthy individuals and found an exquisite coherence at the quantum level. But what kind of light was present in a person who was ill? He tried out his machine on a series of cancer patients. In every instance, the cancer patients had lost these natural periodic rhythms and also their coherence. The lines of internal communication were scrambled. They had lost their connection with the world. In effect, their light was going out.

Just the opposite occurred with multiple sclerosis: MS was a state of too much order. Individuals with this disease were taking in too much light, and this was inhibiting the ability of cells to do their job. Too much cooperative harmony prevented flexibility and individuality: it is like too many soldiers marching in step when they cross a bridge, causing it to collapse. Perfect coherence is an optimum state just between chaos and order. With too much cooperativity, it was as though individual members of the orchestra were no longer able to improvise. MS patients were drowning in light.[23]

Popp also examined the effect of stress. In a stressed state, the rate of biophoton emissions went up – a defense mechanism designed to try to return the patient to equilibrium.

All of these phenomena led Popp to think of biophoton emissions as a sort of correction by a living system of Zero Point Field fluctuations. Every system likes to achieve a minimum of free energy. In a perfect world, all waves would cancel each other out by destructive interference. However, this is impossible with the Zero Point Field, where these tiny fluctuations of energy constantly disturb the system. Emitting photons is a compensatory gesture, to stop this disturbance and attempt a sort of energy equilibrium. As Popp thought of it, the Zero Point Field forces a human being to be a candle. The healthiest body would have the lowest light and be closest to zero state, the most desirable state – the closest living things could get to nothingness.

Popp now recognized that what he'd been experimenting with was even more than a cure for cancer or *gestaltbildung*. Here was a model which provided a better explanation than the current neo-Darwinist theory for how all living things evolve on the planet. Rather than a system of fortunate but ultimately random error, if DNA uses frequencies of all variety as an information tool, this would suggest instead a feedback system of perfect communication through waves which encode and transfer information.

It might also account for the body's capacity for regeneration. The bodies of numerous species of animals have demonstrated the ability to regenerate a lost limb. Experiments with salamanders as far back as the 1930s have shown that an entire limb, a jaw, even the lens of an eye could be amputated but entirely regenerate as though a hidden blueprint were being followed.

This model might also account for the phenomenon of phantom limbs, the strong physical sense among amputees that a missing arm or leg is still present. Many amputees who complain of utterly realistic cramps, aches or tinglings in the missing limb may be experiencing a true physicality which still exists – a shadow of the limb as imprinted on the Zero Point Field.[24]

Popp came to realize that light in the body might even hold the key to health and illness. In one experiment he compared the light emitted from free-range eggs to those produced by battery hens. The photons in the eggs produced by the free-range chickens were far more coherent than those in the battery eggs. He went on to use biophoton emissions as a tool for measuring the quality of food. The healthiest food had the lowest and most coherent intensity of light. Any disturbance in the system would increase the production of photons. Health was a state of perfect subatomic communication, and ill health was a state where communication breaks down. We are ill when our waves are out of synch.

Once Popp began publishing his findings, he began to attract the enmity of the scientific community. Many of his fellow German scientists believed that Popp's bright spark had finally gone out. At his university, students wanting to study biophoton emissions began to be censured. By 1980, when Popp's contract as an assistant professor was finished, the university had an excuse to ask him to leave. Two days before the end of his term, university officials marched into his laboratory and demanded that he surrender all his equipment. Fortunately, Popp had been tipped off about the raid and had hidden his photomultiplier in the basement of the lodgings of a sympathetic student. When he left campus, he left with his precious equipment intact.

Popp's treatment at the hands of the University of Marburg resembled that of a criminal without a fair trial. As an assistant professor of some years standing, Popp was entitled to substantial compensation for his years of service, but the university refused to pay him. He had to sue the university to get the 40,000 marks that were due him. He won his money, but his career lay in ashes. He was a married man with three young children and no apparent means of employment. No university at the time was prepared to touch him.

It looked as though Popp's academic career was finished. He spent two years in private industry with Roedler, a pharmaceutical manufacturer of homeopathic remedies, one of the few organisations to entertain his wild theories. Nevertheless, Popp, a stubborn autocrat in his own labs, was equally stubborn in persisting with his work, convinced of its validity. Eventually, he gained a patron in Professor Walter Nagl of the University of Kaiserslautern, who asked Popp to work with him. Once again, Popp's research caused a revolt among the faculty, who demanded his resignation on the grounds that his work was sullying the university's reputation.

Eventually Popp gained employment from the Technology Center in Kaiserslautern, which is largely sponsored by government grants for application research. It would take some 25 years for him to gather converts from among the scientific community. Slowly a few select scientists from around the globe began to consider that the body's communication system might be a complex network of resonance and frequency. Eventually they would form the International Institute of Biophysics, composed of fifteen groups of scientists from international centres all around the world. Popp had found offices for his new group in Neuss, near Düsseldorf. The brother of a Nobel prizewinner, the grandson of Alexander Gurwitsch, a nuclear physicist from Boston University and nuclear research laboratory CERN in Geneva, two Chinese biophysicists – noted scientists from around the globe at last were beginning to agree with him. Popp's fortunes were beginning to turn. Suddenly he was receiving offers and contracts for professorships from reputable universities around the world.

Popp and his new colleagues went on to study the light emissions of several organisms of the same species, first with an experiment with a type of water flea called *Daphnia*. What they found was nothing short of astonishing. Tests with a photomultiplier showed that the water fleas were sucking up the light emitted from each other. Popp tried the same experiment on small fish and found that they were doing the same. According to his photomultiplier, sunflowers were like a biological vacuum cleaner, moving in the direction of the most solar photons in order to hoover them up. Even bacteria would swallow photons from the medium they had been placed in.[25]

It began to dawn on Popp that these emissions had a purpose outside the body. Wave resonance wasn't simply being used to communicate inside the body, but between living things. Two healthy beings were engaged in 'photon sucking', as he called it, by exchanging photons. Popp realized that this exchange might unlock the secret of some of the animal kingdom's most persistent conundrums: how schools of fish or flocks of birds create perfect and instantaneous coordination. Many experiments on the homing ability of animals demonstrate that it has nothing to do with following habitual trails or scents or even the electromagnetic fields of the earth, but some silent communication, acting like an invisible rubber band, even when animals are separated by miles from humans.[26] For humans there was another possibility. If we could take in the photons of other living things, we also might be able to use the information from them to correct our own light if it went awry.

Popp had begun experimenting with such an idea. If some cancer-causing chemicals could alter the body's biophoton emissions, then it might be the case that other substances could reintroduce better communication. Popp wondered whether certain plant extracts could change the character of biophoton emissions of cancer cells, so that they would began to communicate again with the rest of the body. He began experimenting with a number of non-toxic substances purported to be successful in treating cancer. In all but one instance, the substances only increased the photons from tumor cells, making it even more deadly to the body. The single success story was mistletoe, which seemed to help the body to 'resocialize' the photon emission of tumor cells back to normal. In one of numerous cases, Popp came across a woman in her thirties with breast and vaginal cancer. Popp tried mistletoe and other plant extracts on samples of her cancerous tissue and found that one particular mistletoe remedy created coherence in the tissue similar to that of the body. With the agreement of her doctor, the woman began forgoing any treatment other than this mistletoe extract. After a year, all her laboratory tests were virtually back to normal. A woman who was given up as a terminal cancer case had her proper light restored, just by taking a herb.[27]

To Fritz-Albert Popp, homeopathy was another example of photon sucking. He had begun to think of it as a 'resonance absorber'. Homeopathy rests upon the notion that like is treated with like. A plant extract that at full strength can cause hives in the body is used in an extremely dilute form to cure them. If a rogue frequency in the body could produce certain symptoms, it followed that the high dilution of a substance which would produce the same symptoms would still carry those oscillations. Like a tuning fork in resonance, a suitable homeopathic solution might attract and then absorb the wrong oscillations, allowing the body to return to normal.

Popp thought that electromagnetic molecular signalling might even explain acupuncture. According to the theory of traditional Chinese medicine, the human body has a meridian system running deep in the tissues of the body through which flows an invisible energy which the Chinese term 'the *qi*, or life force. The *qi* supposedly enters the body through these acupuncture points and flows to deeper organ structures (which do not correspond to those of Western human biology), providing energy (and thus the life force). Illness occurs when there is a blockage of this energy anywhere along the pathways. According to Popp, the meridian system may work like wave guides transmitting particular bodily energy to specific zones.

Scientific studies show that many acupuncture points on the body have a dramatically decreased electrical resistance compared with points on the skin surrounding it (10 kilo-ohms at the center of a point, compared with 3 mega-ohms in the surrounding skin).[28] Research has also shown that painkilling endorphins and the steroid cortisol are released through the body when the points are stimulated at low frequency, and important mood-regulating neurotransmitters like serotonin and norepinephrine, at high frequency. The same doesn't occur when the skin surrounding these points is stimulated.[29] Yet other research has proved that acupuncture can cause blood vessels to dilate and increase blood flow to distant organs in the body.[30] Other research demonstrates the existence of meridians as well as the effectiveness of acupuncture for a variety of conditions. Orthopaedic surgeon Dr Robert Becker, who performed a great deal of research on electromagnetic fields in the body, designed a special electrode recording device which would roll along the body like a pizza cutter. After many studies it showed up electrical charges on the same places on every one of the people tested, all corresponding to Chinese meridian points.[31]

There were many possibilities to explore, some of which might pan out, and some not. But Popp was convinced of one thing: his theory of DNA and biophoton emission was correct and this drove the processes of the body. There was no doubt in his mind that biology was driven by the quantum process he'd observed. All he needed were other scientists with experimental evidence to show how it might be so.

CHAPTER FOUR

The Language of the Cell

IN A WHITE PORTAKABIN in Clamart, in the unfashionable outskirts of Paris, a tiny heart, propped atop a bit of purpose-built scaffolding, carried on beating. It was being kept alive courtesy of a small team of French scientists, who administered the right combination of oxygen and carbon dioxide, part of the type of state-of-the-art surgical technique used for heart transplants. In this instance, there was no donor or recipient; the heart had long been divested of its owner, a prime male Hartley guinea pig, and the scientists were only interested in the organ itself and how it was about to react. They'd applied acetylcholine and histamine, two known vasodilators, then atropine and mepyramine, both agonists to the others, and finally measured coronary flow, plus such mechanical changes as beat rate.

There were no surprises here. As expected, the histamine and acetylcholine produced increased blood flow in the coronary arteries, while the mepyramine and atropine inhibited it. The only unusual aspect of the experiment was that the agents of change weren't actually pharmacological chemicals but low-frequency waves of the electromagnetic signals of the cells recorded using a purpose-designed transducer and a computer equipped with a sound card. It was these signals, which take the form of electromagnetic radiation of less than 20 kilohertz, which were applied to the guinea pig heart, and were responsible for speeding it up, just as the chemicals themselves would.[1]

The signal effectively could take the place of the chemicals, for the signal *is* the molecule's signature. The scientific team, which had successfully substituted it for the original, were quietly aware of the explosive nature of their achievement. Through their efforts, the usual theories of molecular signaling and how cells 'talk' to each other had been profoundly modified. They were beginning to demonstrate in the laboratory what Popp had just proposed – that each molecule in the universe had a unique frequency and the language it used to speak to the world was a resonating wave.

As Popp was pondering the larger implications of biophoton emissions, a French scientist had been examining the reverse: the effect of this light on individual molecules. Popp believed that biophoton emissions orchestrated all bodily processes, and the French scientist was finding out the exquisite way in which it worked. The biophoton vibrations Popp had observed in the body caused molecules to vibrate and create their own signature frequency, which acted as its unique driving force and also its means of communication. The French scientist had paused to listen to these tiny oscillations and heard the symphony of the universe. Every molecule of our bodies was playing a note that was being heard round the world.

This discovery represented a permanent and arduous detour in the career of French scientist Jacques Benveniste, which had, up until the 1980s, followed a distinguished, predictable arc. Benveniste, a doctor of medicine, had put in his residency in the Paris hospital system, and then moved into research into allergies, becoming a specialist in the mechanisms of allergy and inflammation. He'd been appointed research director at the French National Institute for Health and Medical Research (INSERM) and distinguished himself by discovering PAF, or platelet activating factor, which is involved in the mechanism of allergies such as asthma.

At 50, Benveniste had the world at his feet. There was no doubt that he would look forward to international acclaim among the establishment. He was proud of being French in a field not necessarily well represented by his countrymen since Descartes. Rumours abounded about the possibility that Benveniste would be one of the few French biologists to be considered as a possible recipient for the Nobel prize. His papers were among those most often cited by scientists at INSERM, a measure of distinction and standing. He'd even received the Silver Medal from CNRS, one of the most prestigious French scientific honors. Benveniste possessed craggy good looks, a regal bearing, and a rakish sense of humor, and he'd been married for 30 years. Nevertheless, neither his marital status nor his present contentment in the slightest curbed a tendency to innocently flirt, an attribute that, as a Frenchman, he considered more or less mandatory.

And then, in 1984, this bright and assured future was accidentally derailed by what turned out to be a small error in computation. Benveniste's laboratory at INSERM had been studying basophil degranulation – the reaction of certain white blood cells to allergens. One day, Elisabeth Davenas, one of his best laboratory technicians, came to him and reported that she'd seen and recorded a reaction in the white blood cells, even though there had been too few molecules of the allergen in the solution.

This had all come about as the result of a simple error in calculation. She had thought the starting solution was more concentrated than it was. In diluting it to what she thought was the usual concentration, she had inadvertently diluted the solution to the point where very few of the original antigen molecules remained.

After examining the data, Jacques virtually shooed her out of his office. The results you are claiming are impossible, he declared, because there are no molecules here.

'You have been experimenting with water,' he told her. 'Go back and do the work over.'

It was only when she tried to repeat the experiment with the same dilution and came up with the same results that he realized that Elisabeth, a meticulous worker, might have stumbled onto something worth investigating. For several weeks, Elisabeth kept returning to his office with the same inexplicable data, showing powerful biological effects from a solution so weakened that it couldn't have enough of the antigen to have caused them, and Jacques attempted to come up with ever more far-fetched explanations to fit these results to some recognizable biological theory. Perhaps it was the presence of a second antibody reacting later, or maybe the reaction to an undisclosed second antigen, he thought. After observing these results, one of the tutors in his laboratory, a doctor who was also a homeopath, happened to remark that these experiments were quite similar to the principle of homeopathy. In that system of medicine, solutions of active substance are diluted to the point where there is virtually none of the original substance left, only its 'memory'. At the time, Jacques didn't even know what homeopathy was – that's how classical a doctor he was – but the research scientist in him had had his appetite sufficiently whetted. He asked Elisabeth to dilute the solutions even more, so that absolutely none of the original active substance remained. In these new studies, no matter how dilute the solution, which was, by now, just plain water, Elisabeth kept getting consistent results, as if the active ingredient were still there.

Because of his background as an allergy specialist, Jacques had used a standard allergy test for his studies, the purpose of which was to effect a typical allergic response in human cells. He isolated basophils, a type of white blood cell which contains antibodies of immunoglobulin E (IgE) type on its surface. It is these cells which are responsible for hypersensitivity reactions in people with allergies.

Jacques chose IgE cells because they easily respond to allergens such as pollen or dust mites, releasing histamine from their intracellular granules, and also to certain anti-IgE antibodies. If this kind of a cell is affected by something, you're not likely to miss it. Another advantage of the IgE is that he could test their staining properties through a test he'd developed and patented at INSERM. Because basophils, like most cells, have a jelly-like appearance, when you're studying them at a lab, you need to stain them in order to see them. But staining, even with a standard dye such as toluidine blue, is subject to change, depending upon many factors – the health of the host, say, and the influence of other cells upon the original. When these IgE cells are exposed to anti-IgE antibodies, it changes their ability to absorb the dye. Anti-IgE has been referred to as a kind of 'biological paint-stripper'[2] because its ability to inhibit the dye is so effective that it can virtually render the basophils invisible again.

The final logic in Benveniste's choice of anti-IgE had to do with the fact that these particular molecules are especially big. If you are attempting to see if water retained its effect even when all anti-IgE molecules had been filtered out of it, there would be no chance that any of them might be accidentally left behind.

In the studies, conducted over four years between 1985 and 1989, and painstakingly recorded in the laboratory books of Elisabeth Davenas, Benveniste's team created high dilutions of the anti-IgE by pouring one-tenth of the previous solution into the next tube and filling it up by adding nine parts of a standard solvent. Each dilution was then vigorously shaken (or successed, as it is technically known), as it is in homeopathic preparations. In total, the team used dilutions like these, of one part solution to nine parts solvent, then kept diluting until there was one part of solution to ninety-nine parts solvent and even one part solution to nine hundred and ninety-nine parts solvent.

Each one of the high dilutions was successively added to the basophils, which were then counted under the microscope. To Jacques' surprise, as much as anyone's, they discovered that they were recording effects in inhibiting dye absorption by up to 66 per cent, even with dilutions watered down to one part in 10^{60}. In later experiments, when the dilutions were serially diluted a hundred-fold, eventually to one part in 10^{120}, where there was virtually no possibility that a single molecule of the IgE was left, the basophils were still affected.

The most unexpected phenomenon was yet to come. Although the potency of the anti-IgE was at its highest at concentrations of one part in 1000 (the third decimal dilution) and then started to decrease with each successive dilution, as you might logically expect, the experiment took a U-turn at the ninth dilution. The effect of the highly dilute IgE began increasing at this point and continued to increase, the more it was diluted.[3] As homeopathy had always claimed, the weaker the solution, the more powerful its effect.

Benveniste joined forces with five different laboratories in four countries, France, Israel, Italy and Canada, all of whom were able to replicate his results. The thirteen scientists then jointly published the results of their four-year collaboration in a 1988 edition of the highly prestigious *Nature* magazine, showing that if solutions of antibodies were diluted repeatedly until they no longer contained a single molecule of the antibody, they still produced a response from immune cells.[4] The authors concluded that none of the molecules they'd started with were present in certain dilutions and that:

> **specific information must have been transmitted during the dilution/shaking process. Water could act as a template for the molecule, for example, by an infinite hydrogen-bonded network, or electric and magnetic fields . . . The precise nature of this phenomenon remains unexplained.**

To the popular press, which pounced on the published paper, Benveniste had discovered 'the memory of water', and his studies were widely regarded as making a valid case for homeopathy. Benveniste himself realized that his results had repercussions far beyond any theory of alternative medicine. If water were able to imprint and store information from molecules, this would have an impact on our understanding of molecules and how they 'talk' to one another in our bodies, as molecules in human cells, of course, are surrounded by water. In any living cell, there are ten thousand molecules of water for each molecule of protein.

Nature also undoubtedly understood the possible repercussions of this finding on the accepted laws of biochemistry. The editor, John Maddox, had consented to publish the article, but he did so after taking an unprecedented step – placing an editorial addendum at the bottom of the article:

> *Editorial reservation*
> Readers of this article may share the incredulity of the many referees who have commented on several versions of it during the past several months. The essence of the result is that an aqueous solution of an antibody retains its ability to evoke a biological response even when diluted to such an extent that there is a negligible chance of their being a single molecule in any sample. There is no physical basis for such an activity. With the kind collaboration of Professor Benveniste, *Nature* has therefore arranged for independent investigators to observe repetitions of the experiments. A report of this investigation will appear shortly.

In his own editorial, Maddox also invited readers to pick holes in the Benveniste study.[5]

Benveniste was a proud man, not afraid to wave a fist in the face of the Establishment. He was not only willing to stick his head above the parapet in choosing to publish in one of the most conservative journals in the whole of the scientific community, but then, when they doubted him, he eagerly snatched up the gauntlet they'd thrown down by agreeing to their request to reproduce his results at his laboratory.

Four days after publication, Maddox himself arrived with what Benveniste described as a scientific 'fraud squad', composed of Walter Stewart, a well-known quackbuster, and James Randi, a professional magician who tended to be called in to expose scientific work that had actually been arrived at by sleight of hand. Were a magician, a journalist and a quackbuster the best possible team to assess the subtle changes in biological experimentation, wondered Benveniste. Under their watchful eye, Elisabeth Davenas performed four experiments, one blinded, all of which, Benveniste said, were successful. Nevertheless, Maddox and his team disputed the findings and decided to change the experimental protocol and tighten the coding procedures, even, in a melodramatic gesture, taping the code to the ceiling. Stewart insisted on carrying out some of the experiments himself and changed some of their design even though, Benveniste claimed, he was untrained in these particular experiments.

Under their new protocol, and amid a charged atmosphere implying that the INSERM team were hiding something, three more tests were done and shown not to work. At this point, Maddox and his team had their results and promptly left, first asking for photocopies of 1500 of Benveniste's papers.

Soon after their five-day visit, *Nature* published a report entitled 'High dilution experiments a delusion'. It claimed that Benveniste's lab had not observed good scientific protocol. It discounted supporting data from other labs. Maddox expressed surprise that the studies didn't work all the time, when this is standard in biological studies – one reason Benveniste had conducted more than 300 trials before publishing. The Maddox judgment also failed to note that the staining test is highly sensitive and can be tipped with the slightest change in experimental condition, so that some donor blood isn't affected by even high concentrations of anti-IgE. They expressed dismay that two of Benveniste's co-authors were being funded by a manufacturer of homeopathic medicines. Industry funding is standard in scientific research, countered Benveniste. Were they implying that the results were altered to please the sponsor?

Benveniste fought back with an impassioned response and a plea for scientific open-mindedness:

> Salem witchhunts or McCarthy-like prosecutions will kill science. Science flourishes only in freedom . . . The only way definitively to establish conflicting results is to reproduce them. It may be that all of us are wrong in good faith. This is no crime but science as usual.[6]

Nature's results had a devastating effect upon Benveniste's reputation and his position at INSERM. A scientific council of INSERM censured his work, claiming in near unanimous statements that he should have performed other experiments 'before asserting that certain phenomena have escaped two hundred years of chemical research.'[7] INSERM refused to listen to Benveniste's objections about the quality of the *Nature* investigation and prevented him from continuing. Rumours circulated about mental imbalance and fraud. Letters poured in to *Nature* and other publications, calling his work 'dubious science', a 'cruel hoax' and 'pseudo-science'.[8]

Benveniste was given several chances to gracefully bow out of this work and no professional reason to continue to pursue it. By standing by his original work, he was certain to destroy the career he'd been building. Benveniste had got to the top of his position at INSERM and had no desire to be director. He'd never had ambition for a career, but only wished to carry on with his research. By that time, he also felt he had no choice – the genie was already out of the bottle. He had uncovered evidence that demolished everything he had been taught to believe about cell communication, and there was now no turning back. But also there was the undeniable thrill of it. Here was the most compelling research he could think of, the most explosive of results he could imagine. This was like, as he enjoyed putting it, peering under the skirt of nature. Benveniste left INSERM, and sought support from private sources such as DigiBio, which enabled him and Didier Guillonnet, a gifted engineer from École Centrale Paris, who joined him in 1997, to carry on their work. After the *Nature* fiasco, they moved on to 'digital biology', a discovery they made not in a single moment of inspiration, but after eight years of following a logical trail of cautious experimentation.[9]

The memory of water studies had prompted Benveniste to examine the manner in which molecules communicate within a living cell. In all aspects of life, molecules must speak to each other. If you are excited, your adrenals pump out more adrenaline, which must tell specific receptors to get your heart to beat faster. The usual theory, called the Quantitative Structure-Activity Relationship (QSAR), is that two molecules that match each other structurally exchange specific (chemical) information, which occurs when they bump into each other. It's rather like a key finding its own keyhole (which is why this theory is often also called the key–keyhole, or lock-and-key interaction model). Biologists still adhere to the mechanistic notions of Descartes that there can only be reaction through contact, some sort of impulsive force. Although they accept gravity, they reject any other notions of action at a distance.

If these occurrences are due to chance, there's very little statistical hope of their happening, considering the universe of the cell. In the average cell, which contains one molecule of protein for every ten thousand molecules of water, molecules jostle around the cell like a handful of tennis balls floating about in a swimming pool. The central problem with the current theory is that it is too dependent upon chance and also requires a good deal of time. It can't begin to account for the speed of biological processes, like anger, joy, sadness or fear. But if instead each molecule has its own signature frequency, its receptor or molecule with the matching spectrum of features would tune into this frequency, much as your radio tunes into a specific station, even over vast distances, or one tuning fork causes another tuning fork to oscillate at the same frequency. They get in resonance – the vibration of one body is reinforced by the vibration of another body at or near its frequency. As these two molecules resonate on the same wavelength, they would then begin to resonate with the next molecules in the biochemical reaction, thus creating, in Benveniste's words, a 'cascade' of electromagnetic impulses travelling at the speed of light. This, rather than accidental collision, would better explain how you initiate a virtually instantaneous chain reaction in biochemistry. It also is a logical extension of the work of Fritz Popp. If photons in the body excite molecules along the entire spectrum of electromagnetic frequencies, it is logical that they would have their own signature frequency.

Benveniste's experiments decisively demonstrated that cells don't rely on the happenstance of collision but on electromagnetic signalling at low frequency (less than 20 kHz) electromagnetic waves. The electromagnetic frequencies that Benveniste has studied correspond with frequencies in the audio range, even though they don't emit any actual noise that we can detect. All sounds on our planet – the sound of water rippling in a stream, a crack of thunder, a shot fired, a bird chirping – occur at low frequency, between 20 hertz and 20 kilohertz, the range in which the human ear can hear.

According to Benveniste's theory, two molecules are then tuned into each other, even at long distance, and resonate to the same frequency. These two resonating molecules would then create another frequency, which would then resonate with the next molecule or group of molecules, in the next stage of the biological reaction. This would explain, in Benveniste's view, why tiny changes in a molecule – the switching of a peptide, for example – would have a radical effect on what that molecule actually does.

This is not so farfetched, considering what we already know about how molecules vibrate. Both specific molecules and intermolecular bonds emit certain specific frequencies which can be detected billions of light-years away, through the most sensitive of modern telescopes. These frequencies have long been accepted by physicists, but no one in the biological community save Fritz-Albert Popp and his predecessors has paused to consider whether they actually have some purpose. Others before Benveniste, such as Robert O. Becker and Cyril Smith, had conducted extensive experimentation on electromagnetic frequencies in living things. Benveniste's contribution was to show that molecules and atoms had their own unique frequencies by using modern technology both to record this frequency and to use the recording itself for cellular communication.

From 1991, Benveniste demonstrated that you could transfer specific molecular signals simply by using an amplifier and electromagnetic coils. Four years later, he was able to record and replay these signals using a multimedia computer. Over thousands of experiments, Benveniste and Guillonnet recorded the activity of the molecule on a computer and replayed it to a biological system ordinarily sensitive to that substance. In every instance, the biological system has been fooled into thinking it has been interacting with the substance itself and acted accordingly, initiating the biological chain reaction, just as it would if in the actual presence of the genuine molecule.[10] Other studies have also shown that Benveniste's team could erase these signals and stop activity in the cells through an alternating magnetic field, work they performed in collaboration with Centre National de la Recherche Scientifique in Medudon, France. The inescapable conclusion: as Fritz-Albert Popp theorized, molecules speak to each other in oscillating frequencies. It appeared that the Zero Point Field creates a medium enabling the molecules to speak to each other nonlocally and virtually instantaneously.

The DigiBio team tested out digital biology on five types of studies: basophilic activation; neutrophilic activation; skin testing; oxygen activity; and, most recently, plasma coagulation. Like whole blood, plasma, the yellowy liquid of the blood, which carries protein and waste products, will coagulate. To control for that ability, you must first remove the calcium in the plasma, by chelating – chemically grabbing – it. If you then add water with calcium to the blood, it will coagulate, or clot. Adding heparin, a classic anti-coagulant drug, will prevent the blood from clotting, even in the presence of the calcium.

In Benveniste's most recent study, he took a test-tube of this plasma with calcium chelated out, then added water containing calcium which has been exposed to the 'sound' of heparin transmitted via the signature digitized electromagnetic frequency. As with all his other experiments, the signature frequency of heparin works as though the molecules of heparin itself were there: in its presence, the blood is more reluctant than usual to coagulate.

In perhaps the most dramatic of his experiments, Benveniste showed that the signal could be sent across the world by email or mailed on a floppy disk. Colleagues of his at Northwestern University in Chicago recorded signals from ovalbumin (Ova), acetylcholine (Ach), dextran and water. The signals from the molecules were recorded on a purpose-designed transducer and a computer equipped with a sound card. The signal was then recorded on a floppy disk and sent by regular mail to the DigiBio Laboratory in Clamart. In later experiments, the signals were also sent by email as attached documents. The Clamart team then exposed ordinary water to the signals of this digital Ova or Ach or ordinary water and infused either the exposed water or the ordinary water to isolated guinea pig hearts. All the digitised water produced highly significant changes in coronary flow, compared with the controls – which just contained ordinary, non-exposed water. The effects from the digitized water were identical to effects produced on the heart by the actual substances themselves.[11]

Giuliano Preparata and his colleague Emilio Del Giudice, two Italian physicists at the Milan Institute for Nuclear Physics, were working on a particularly ambitious project – to explain why certain matter in the world stays in one piece. Scientists understand gases to a large extent through the laws of classical physics, but are still largely ignorant of the actual workings of liquids and solids – that is, any sort of condensed matter. Gases are easy because they consist of individual atoms or molecules which behave individually in large spaces. Where scientists have trouble is with atoms or molecules packed tightly together and how they behave as a group. Any physicist is at a loss to tell you why water doesn't just evaporate into gas or why atoms in a chair or a tree stay that way, particularly if they are only supposed to communicate with their most immediate neighbor and be held together by short-range forces.[12]

Water is among the most mysterious of substances, because it is a compound formed from two gases, yet it is liquid at normal temperatures and pressures. In their studies, Del Giudice and Preparata have demonstrated mathematically that when closely packed together, atoms and molecules exhibit a collective behavior, forming what they have termed 'coherent domains'. They are particularly interested in this phenomenon as it occurs in water. In a paper published in *Physical Review Letters*, Preparata and Del Giudice demonstrated that water molecules create coherent domains, much as a laser does. Light is normally composed of photons of many wavelengths, like colors in a rainbow, but photons in a laser have a high degree of coherence, a situation akin to a single coherent wave, like one intense color.[13] These single wavelengths of water molecules appear to become 'informed' in the presence of other molecules – that is, they tend to polarize around any charged molecule – storing and carrying its frequency so that it may be read at a distance. This would mean that water is like a tape recorder, imprinting and carrying information whether the original molecule is still there or not. The shaking of the containers, as is done in homeopathy, appears to act as a method of speeding up this process.[14] So vital is water to the transmission of energy and information that Benveniste's own studies actually demonstrate that molecular signals cannot be transmitted in the body unless you do so in the medium of water.[15] In Japan, a physicist called Kunio Yasue of the Research Institute for Information and Science, Notre Dame Seishin University in Okayama, also found that water molecules have some role to play in organizing discordant energy into coherent photons – a process called 'superradiance'.[16]

This suggests that water, as the natural medium of all cells, acts as the essential conductor of a molecule's signature frequency in all biological processes and that water molecules organize themselves to form a pattern on which can be imprinted wave information. If Benveniste is right, water not only sends the signal but also amplifies it.

The most important aspect of scientific innovation is not necessarily the original discovery, but the people who copy the work. It is only the replication of initial data that legitimizes your research and convinces the orthodox scientific community that you might be onto something. Despite the virtually universal derision of Benveniste's results by the Establishment, reputable research slowly began to appear elsewhere. In 1992, FASEB (the Federation of American Societies for Experimental Biology) held a symposium, organized by the International Society for Bioelectricity, examining the interactions of electromagnetic fields with biological systems.[17] Numerous other scientists have replicated high-dilution experiments,[18] and several others have endorsed and successfully repeated experiments using digitized information for molecular communication.[19] Benveniste's latest studies were replicated eighteen times in an independent lab in Lyon, France, and in three other independent centres.

Several years after the memory of water *Nature* episode, scientific teams still tried to prove Benveniste wrong. Professor Madelene Ennis of Queen's University in Belfast joined a large pan-European research team, with hopes of showing, once and for all, that homeopathy and water memory were utter nonsense. A consortium of four independent laboratories in Italy, France, Belgium and Holland, led by Professor M. Roberfroid of the

Catholic University of Louvain, in Brussels, carried out a variation of Benveniste's original experiment with basophil degranulation. The experiment was impeccable. None of the researchers knew which was the homeopathic solution and which pure water. All the solutions had even been prepared by labs which had nothing further to do with the trial. Results were also coded and decoded and tabulated by an independent researcher also unconnected with the study.

In the end, three of four labs got statistically significant results with the homeopathic preparations. Professor Ennis still didn't believe these results and put them down to human error. To eliminate the possible vagaries of humans, she applied an automated counting protocol to the figures she had. Nevertheless, even the automated results showed the same. The high dilutions of the active ingredient worked, whether the active ingredient was actually present or water so dilute that none of the original substance remained. Ennis was forced to concede: 'The results compel me to suspend my disbelief and to start searching for rational explanations for our findings.'[20]

This represented the last straw to Benveniste. If Ennis's results were negative, they would have been published in *Nature*, thereby forever consigning his work to the trash heap. Because their results agreed with his, they were published in a relatively obscure journal, a few years after the event, a guarantee that no one would really notice.

Besides Ennis's results, there were all the scientific studies of homeopathy which lent support to Benveniste's findings. Excellent, double-blind, placebo-controlled trials showed that homeopathy works for, among many conditions, asthma,[21] diarrhea,[22] upper respiratory tract infections in children[23] and even heart disease.[24] Of at least 105 trials of homeopathy, 81 showed positive results.

The most unassailable were carried out in Glasgow by Dr David Reilly, whose double-blind, placebo-controlled studies showed that homeopathy works for asthma, with all the usual checks and balances of a pristine scientific study.[25] Despite the scientific design of the trial, an editorial in *The Lancet*, redolent of *Nature's* response to Benveniste's initial findings, agreed to publish the results but simply refused to accept them:

> What could be more absurd than the notion that a substance is therapeutically active in dilutions so great that the patient is unlikely to receive a single molecule of it? [said the editorial]. Yes,

> the dilution principle of homeopathy is absurd; so the reason for any therapeutic effect presumably lies elsewhere.[26]

On reading *The Lancet*'s on-going debate on the Reilly studies, Benveniste couldn't resist responding:

> This recalls, inexorably, the wonderfully self-sufficient contribution of a nineteenth-century French academician to the heated debate over the existence of meteorites, which animated the scientific community at the time: 'Stones do not fall from the sky because there are no stones in the sky.'[27]

Benveniste was so tired of laboratories trying and sometimes failing to replicate his work that he had Guillonnet build him a robot. Nothing much more than a box with an arm which moves in three directions, the robot could handle everything but the initial measuring. All one had to do was to hand it the bare ingredients plus a bit of plastic tubing, push the button and leave. The robot would take the water containing calcium, place it into a coil, play the heparin signal for five minutes, so that the water is 'informed', then mix the informed water in its test-tube with the plasma, put the mixture in a measuring device, read the results and offer them up to whoever is doing the investigation. Benveniste and his team carried out hundreds of experiments using their robot, but the main idea was to hand out a batch of these devices to other labs. In this way, both the other centres and the Clamart team can ensure that the experiment is universally standardized and an identical protocol carried out correctly.

While working with his robot, Benveniste discovered on a large scale what Popp had witnessed in the laboratory with his water fleas – evidence that the electromagnetic waves from living things were having an effect on their environment.

Once Benveniste had got his robot up and working, he discovered that generally it worked well, except for certain occasions. Those occasions were always the days when a particular woman was present in the lab. *Cherchez la femme*, Benveniste thought, although in the Lyon lab, which was replicating their results, a similar situation occurred, this time with a man. In his own lab, Benveniste conducted several experiments, by hand and by robot, to isolate what it was the woman was doing which prevented the experiment from working. Her scientific method was impeccable and she followed the protocol to the letter. The woman herself, a doctor and biologist, was an experienced, meticulous worker. Nevertheless, on no occasion did she get any results. After six months of such studies there was only a single conclusion: something about her very presence was preventing a positive result.

It was vital that he got to the nub of the problem, for Jacques knew what was at stake. He might send his robot to a laboratory in Cambridge, and if they got poor results as a result of a particular person, the lab would conclude that the experiment itself was at fault, when the problem had to do with something or someone in the environment.

There is nothing subtle about biological effects. Change the structure or shape of a molecule only slightly and you will completely alter the ability of the molecule to slot in with its receptor cells. On or off, success or failure. A drug works or it doesn't. In this case, something in the woman in question was completely interfering with the communication of cells in his experiment.

Benveniste suspected that the woman must be emitting some form of waves that were blocking the signals. Through his work he developed a means of testing for these, and he soon discovered that she was emitting electromagnetic fields which were interfering with the communication signalling of his experiment. Like Popp's carcinogenic substances, she was a frequency scrambler. This seemed too incredible to believe – more the realm of witchcraft than science, Benveniste thought. He then had the particular woman hold a tube of homeopathic granules in her hand for five minutes, and then tested the tube with his equipment. All activity – all molecular signaling – had been erased.[28]

Benveniste wasn't a theorist. He wasn't even a physicist. He'd accidentally trespassed into the world of electromagnetism and now was stuck here, experimenting in what for him was completely foreign territory – the memory of water and the ability of molecules to vibrate at very high and very low frequencies. These were the two mysteries that he was getting no closer to solving. All that he could do was to carry on where he felt most comfortable – with his laboratory experiments – showing that these effects were real. But one thing did seem clear to him. For some unknown reason that he didn't dwell upon, these signals also appeared to be sent outside the body and somehow were being taken in and listened to.

CHAPTER FIVE

Resonating with the World

VIRTUALLY EVERY EXPERIMENT HAD been a failure. The rats were not performing as expected. The entire point of the exercise, as far as Karl Lashley was concerned, had been to find where the engrams were – the precise location in the brain where memories were stored. The name 'engram' had been coined by Wilder Penfield in the 1920s after he thought he'd discovered that memories had an exact address in the brain. Penfield had performed extraordinary research on epileptic patients with anaesthetized scalps while they were fully conscious, showing that if he stimulated certain parts of their brains with electrodes, specific scenes from their past could be evoked in living color and excruciating detail. Even more amazingly, whenever he had stimulated the same spot in the brain (often unbeknownst to the patient) it seemed to elicit the same flashback, with the same level of detail.

Penfield, and an army of scientists after him, naturally concluded that certain portions of the brain were allotted to hold captive specific memories. Every last detail of our lives had been carefully encoded in specific spots in the brain, like guests at a restaurant placed at certain tables by a particularly exacting maitre d'. All we needed to find was who was sitting where – and, perhaps as a bonus, who the maitre d' was.

For nearly 30 years Lashley, a renowned American neuropsychologist, had been looking for engrams. It was 1946, and at his laboratory at the Yerkes Laboratory of Primate Biology in Florida, he'd been searching across all sorts of species to find out what it was in the brain – or where it was – that was responsible for memory. He'd thought that he would be amplifying Penfield's findings, when all he seemed to be doing was proving him wrong. Lashley tended to the hypercritical, and small wonder. It was as though his life's entire oeuvre had a singularly negative purpose: to disprove all the work of his forebears. The other gospel of the time that still held the scientific community in thrall, but which Lashley was busily disproving, was the notion that every psychological process had a measurable physical manifestation – the move of a muscle, the secretion of a chemical. Once again, the brain was simply, fussily, the maitre d'.

Although he'd mainly been working in primate research in his early work, he'd then moved onto rats. He'd built them a jumping stand, where they learned to jump through miniature doors to reach a reward of food. To underscore the object of the exercise, those that didn't respond correctly fell into pond water.[1]

Once he was convinced that they'd learned the routine, Lashley systematically set about trying to surgically blot out that memory. For all his criticism of the failings of other researchers, Lashley's own surgical technique was a mess – a makeshift and hurried operation. His was a laboratory protocol that would have incensed any modern-day animal-rights champion. Lashley didn't employ aseptic technique, largely because it wasn't considered necessary for rats. He was a crude and sloppy surgeon, by any medical standard, possibly deliberately so, sewing up wounds with a simple stitch – a perfect recipe for brain infection in larger mammals – but no cruder than most brain researchers of the day. After all, none of Ivan Pavlov's dogs survived his brain surgery, all succumbing to brain abscesses or epilepsy.[2] Lashley sought to deactivate certain portions of his rats' brains to find which part held the precious key to specific memories. To accomplish this delicate task he chose as his surgical instrument his wife's curling iron – *a curling iron!* – and simply burned off the part he wished to remove.[3]

His initial attempts to find the seat of specific memories failed; the rats, though sometimes even physically impaired, remembered exactly what they'd been taught. Lashley fried more and more sections of brain; the rats still seemed to make it through the jumping stand. Lashley became even more liberal with the curling iron, working through one part of the brain to the next, but still it didn't seem to have any effect on the rat's ability to remember. Even when he'd injured the vast majority of the brains of individual rats – and a curling iron caused much more damage to the brain than any clean surgical cut – their motor skills might be impaired, and they might stagger disjointedly along, *but the rats always remembered the routine.*

Although they represented a failure of sorts, the results appealed to the iconoclast in Lashley. The rats had confirmed what he had long suspected. In his 1929 monograph *Brain Mechanisms and Intelligence*, a small work that had first gained him notoriety with its radical notions, Lashley had already elucidated his view that cortical function appeared to be equally potent everywhere.[4] As he would later point out, the necessary conclusion from all his experimental work 'is that learning just is not possible at all'.[5] When it came to cognition, for all intents and purposes, the brain was a mush.[6]

For Karl Pribram, a young neurosurgeon who'd relocated to Florida just to do research with the great man, Lashley's failures were something of a revelation. Pribram had bought Lashley's monograph for ten cents secondhand, and when he first arrived in Florida, he hadn't been shy about challenging it with the same fervor Lashley had reserved for many of his peers. Lashley had been stimulated by his bright upstart apprentice, whom he would eventually regard as the closest he ever had to a son.

All of Pribram's own views about memory and the brain's higher cognitive processes were being turned on their heads. If there was no one single spot where specific memories were stored – and Lashley had burnt, variously, every part of a rat's brain – then our memories and possibly other higher cognitive processes – indeed, everything that we term 'perception' – must somehow be distributed throughout the brain.

In 1948, Pribram, who was 29 at the time, accepted a position at Yale University, which had the best neuroscience laboratory in the world. His intention was to study the functions of the frontal cortex of monkeys, in an attempt to understand the effects of frontal lobotomies being performed on thousands of patients at the time. Teaching and carrying out research appealed to him far more than the lucrative life of a neurosurgeon; at one point some years later he would turn down a $100,000 salary at New York's Mt Sinai for the relatively impoverished salary of a professor. Like Edgar Mitchell, Pribram always thought of himself as an explorer, rather than a doctor or healer; as an eight-year-old he'd read over and over – at least a dozen times – the exploits of Admiral Byrd in navigating the North Pole. America itself represented a new frontier to conquer for the boy, who'd arrived at that age from Vienna. Pribram was the son of a famous biologist who'd relocated his family to the US in 1927 because he'd felt that Europe, war-torn and impoverished after the First World War, was no place to raise a child. As an adult, possibly because he'd been so slight of build and not really the stuff of hearty physical exploration (in later life he'd resemble an elfin version of Albert Einstein, with the same majestic drapery of white shoulder-length hair) Karl chose the human brain as his exploratory terrain.

After leaving Lashley and Florida, Pribram would spend the next 20 years pondering the mysteries surrounding the organization of the brain, perception and consciousness. He would set up his own experiments on monkeys and cats, painstakingly carrying out systems studies to work out what part of the brain does what. His laboratory was among the first to identify the location of cognitive processes, emotion and motivation, and he was extraordinarily successful. His experiments clearly showed that all these functions had a specific address in the brain – a finding that Lashley was hard-pressed to believe.

What puzzled him most was a fundamental paradox: cognitive processing had very precise locations in the brain, but within these locations, the processing itself seemed to be determined by, as Lashley had put it, 'masses of excitations . . . without regard to particular nerve cells'.[7] It was true that parts of the brain performed specific functions, but the actual processing of the information seemed to be carried out by something more basic than particular neurons – certainly something that was not particular to any group of cells. For instance, storage appeared to be distributed throughout a specific location and sometimes beyond. *But through what mechanism was this possible?*

Like Lashley, much of Pribram's early work on higher perception appeared to contradict the received wisdom of the day. The accepted view of vision – for the most part still accepted today – is that the eye 'sees' by having a photographic image of the scene or object reproduced onto the cortical surface of the brain, the part which receives and interprets vision like an internal movie projector. If this were true, the electrical activity in the visual cortex should mirror precisely what is being viewed – and this is true to some extent at a very gross level. But in a number of experiments, Lashley had discovered that you could sever virtually all of a cat's optic nerve without apparently interfering whatsoever with its ability to see what it was doing. To his astonishment, the cat apparently continued to see every detail as it was able to carry out complicated visual tasks. If there were something like an internal movie screen, it was as though the experimenters had just demolished all but a few inches of the projector, and yet all of the movie was as clear as it had been before.[8]

In other experiments, Pribram and his associates had trained a monkey to press a certain bar if he was shown a card with a circle on it and another bar if shown a card with stripes. Planted in the monkey's visual cortex were electrodes which would register the brain waves when the monkey saw a circle or stripes. What Pribram was testing for was simply to see if the brain waves differed according to the shape on the card.

What he discovered instead was that the monkey's brain not only registered a difference related to the design on the card, but also whether he'd pressed the right bar and even his intention to press the bar before he did. This result convinced Pribram that control was being formulated and sent down from higher areas in the brain to the more primary receiving stations. This must mean that something far more complicated was happening than what was widely believed at the time, which was that we see and respond to outside stimuli through a simple tunnel flow of information, which flows in from our sense organs to the brain and flows out from the brain to our muscles and glands.[9]

Pribram spent a number of years conducting studies measuring the brain activities of monkeys as they performed certain tasks, to see if he could isolate any further the precise location where patterns and colors were being perceived. His studies kept coming up with yet more evidence that brain response was distributed in patches all across the cortex. In another study, this time of newborn cats, which had been given contact lenses with either vertical or horizontal stripes, Pribram's associates found that the behavior of the horizontally oriented cats wasn't markedly different from that of the vertically oriented ones, even though their brain cells were now oriented either horizontally or vertically. This meant that perception couldn't be occurring with line detection.[10] His experiments and those of others like Lashley were at odds with many of the prevailing neural theories of perception. Pribram was convinced that no images were being projected internally and that there must be some other mechanism allowing us to perceive the world as we do.[11]

Pribram had moved from Yale to the Center for Advanced Study in the Behavioral Sciences at Stanford University in 1958. He might never have formulated any alternative view if his friend Jack Hilgard, a noted psychologist at Stanford, hadn't been updating a textbook in 1964 and needed some up-to-date view of perception. The problem was that the old notions about electrical 'image' formation in the brain – the supposed correspondence between images in the world and the brain's electrical firing – had been disproved by Pribram, and his own monkey studies made him extremely dubious about the latest, most popular theory of perception – that we know the world through line detectors. Just to focus on a face would require a new huge computation by the brain anytime you moved a few inches away from it. Hilgard kept pressing him. Pribram hadn't a clue as to what kind of theory he could give his friend, and he kept racking his brain to offer up some positive angle. Then one of his colleagues chanced across an article in *Scientific American* by Sir John Eccles, the noted Australian physiologist, who postulated that imagination might have something to do with microwaves in the brain. Just a week later, another article appeared, written by Emmet Leith, an engineer at the University of Michigan, about split laser beams and optical holography, a new technology.[12]

It had been right there, all along, right in front of his nose. This was just the metaphor he'd been looking for. The concept of wave fronts and holography seemed to hold the answer to questions he'd been posing for 20 years. Lashley himself had formulated a theory of wave interference patterns in the brain but abandoned it because he couldn't envision how they could be generated in the cortex.[13] Eccles' ideas appeared to solve that problem. Pribram now thought that the brain must somehow 'read' information by transforming ordinary images into wave interference patterns, and then transform them again into virtual images, just as a laser hologram is able to. The other mystery solved by the holographic metaphor would be memory. Rather than precisely located anywhere, memory would be distributed everywhere, so that each part contained the whole.

During a UNESCO meeting in Paris, Pribram met up with Dennis Gabor, who'd won the Nobel prize in the 1940s for his discovery of holography in his quest to produce a microscope powerful enough to see an atom. Gabor, the first engineer to win the Nobel prize in physics, had been working over the mathematics of light rays and wavelengths. In the process he'd discovered that if you split a light beam, photograph objects with it and store this information as wave interference patterns, you could get a better image of the whole than you could with the flat two dimensions you get by recording point-to-point intensity, the method used in ordinary photography. For his mathematical calculations, Gabor had used a series of calculus equations called Fourier transforms, named after the French mathematician Jean Fourier, who'd developed it early in the nineteenth century. Fourier first began work on his system of analysis, which has gone on to be an essential tool of modern-day mathematics and computing, when working out, at Napoleon's request, the optimum interval between shots of a cannon so that the barrel wouldn't overheat. Fourier's method was eventually found to be able to break down and precisely describe patterns of any complexity into a mathematical language describing the relationships between quantum waves. Any optical image could be converted into the mathematical equivalent of interference patterns, the information that results when waves superimpose on each other. In this technique, you also transfer something that exists in time and space into 'the spectral domain' – a kind of timeless, spaceless shorthand for the relationship between waves, measured as energy. The other neat trick of the equations is that you can also use them in reverse, to take these components representing the interactions of waves – their frequency, amplitude and phase – and use them to reconstruct any image.[14]

The evening they were together, Pribram and Gabor drank a particularly memorable bottle of Beaujolais and covered three napkins with complicated Fourier equations, to work out how the brain might be capable of managing this intricate task of responding to certain wave-interference patterns and then converting this information into images.[15] There were numerous fine points to be worked out in the laboratory; the theory wasn't complete. But they were convinced of one thing: perception occurred as a result of a complex reading and transforming of information at a different level of reality.

To understand how this is possible, it's useful to understand the special properties of waves, which are best illustrated in a laser optical hologram, the metaphor that so captured Pribram's imagination. In a classic laser hologram, a laser beam is split. One portion is reflected off an object – a china teacup, say – the other is reflected by several mirrors. They are then reunited and captured on a piece of photographic film. The result on the plate – which represents the interference pattern of these waves – resembles nothing more than a set of squiggles or concentric circles.

However, when you shine a light beam from the same kind of laser through the film, what you see is a fully realized, incredibly detailed, three-dimensional virtual image of the china teacup floating in space (an example of this is the image of Princess Leia which gets generated by R2D2 in the first movie of the *Star Wars* series). The mechanism by which this works has to do with the properties of waves that enables them to encode information and also the special quality of a laser beam, which casts a pure light of only a single wavelength, acting as a perfect source to create interference patterns. When your split beams both arrive on the photographic plate, one half provides the patterns of the light source and the other picks up the configuration of the teacup and both together interfere. By shining the same type of light source on the film, you pick up the image that has been imprinted. The other strange property of holography is that each tiny portion of the encoded information contains the whole of the image, so that if you chopped up your photographic plate into tiny pieces, and shone a laser beam on any one of them, you would get a full image of the teacup.

Although the metaphor of the holograph was important to Pribram, the real significance of his discovery was not holography *per se*, which conjures up a mental image of the three-dimensional ghostly projection, or a universe which is only our projection of it. It was the unique ability of quantum waves to store vast quantities of information in a totality and in three dimensions, and for our brains to be able to read this information and from this to create the world. Here was finally a mechanical device that seemed to replicate the way that the brain actually worked: how images were formed, how they were stored and how they could be recalled or associated with something else. Most important, it gave a clue to the biggest mystery of all for Pribram: how you could have localized tasks in the brain but process or store them throughout the larger whole. In a sense, holography is just convenient shorthand for wave interference – the language of The Field.

The final important aspect of Pribram's brain theory, which would come a little later, had to do with another discovery of Gabor. He'd applied the same mathematics used by Heisenberg in quantum physics for communications – to work out the maximum amount that a telephone message could be compressed over the Atlantic cable. Pribram and some of his colleagues went on to develop his hypothesis with a mathematical model demonstrating that this same mathematics also describes the processes of the human brain. He had come up with something so radical that it was almost unthinkable – a hot, living thing like the brain functioned according to the weird world of quantum theory.

When we observe the world, Pribram theorized, we do so on a much deeper level than the sticks-and-stones world 'out there'. Our brain primarily talks to itself and to the rest of the body not with words or images, or even bits or chemical impulses, but in the language of wave interference: the language of phase, amplitude and frequency – the 'spectral domain'. We perceive an object by 'resonating' with it, getting 'in synch' with it. To know the world is literally to be on its wavelength.

Think of your brain as a piano. When we observe something in the world, certain portions of the brain resonate at certain specific frequencies. At any point of attention, our brain presses only certain notes, which trigger strings of a certain length and frequency.[16] This information is then picked by the ordinary electrochemical circuits of the brain, just as the vibrations of the strings eventually resonate through the entire piano.

What had occurred to Pribram is that when we look at something, we don't 'see' the image of it in the back of our heads or on the back of our retinas, but in three dimensions and out in the world. It must be that we are creating and projecting a virtual image of the object out in space, in the same place as the actual object, so that the object and our perception of the object coincide. This would mean that the art of seeing is one of transforming. In a sense, in the act of observation, we are transforming the timeless, spaceless world of interference patterns into the concrete and discrete world of space and time – the world of the very apple you see in front of you. We create space and time on the surface of our retinas. As with a hologram, the lens of the eye picks up certain interference patterns and then converts them into three-dimensional images. It requires this type of virtual projection for you reach out to touch an apple where it really is, not in some place inside your head. If we are projecting images all the time out in space, our image of the world is actually a virtual creation.

According to Pribram's theory, when you first notice something, certain frequencies resonate in the neurons in your brain. These neurons send information about these frequencies to another set of neurons. The second set of neurons makes a Fourier translation of these resonances and sends the resulting information to a third set of neurons, which then begins to construct a pattern that eventually will make up the virtual image you create of the apple out in space, on top of the fruit bowl.[17] This three-fold process makes it far easier for the brain to correlate separate images – which is easily achieved when you are dealing with wave interference shorthand but extremely awkward with an actual real-life image.

After seeing, Pribram reasoned, the brain must then process this information in the shorthand of wave-frequency patterns and scatter these throughout the brain in a distributed network, like a local area network copying all major instructions for many employees in the office. Storing memory in wave interference patterns is remarkably efficient, and would account for the vastness of human memory. Waves can hold unimaginable quantities of data – far more than the 280 quintillion (280,000,000,000,000,000,000) bits of information which supposedly constitute the average human memory accumulated through an average lifespan.[18] It's been said that with holographic wave-interference patterns, all of the US Library of Congress, which contains virtually every book ever published in English, would fit onto a large sugar cube.[19] The holographic model would also account for the instant recall of memory, often as a three-dimensional image.

Pribram's theories about the distributed role of memory and the wave-front language of the brain met with a great deal of disbelief, especially in the 1960s, when they were first published. Chief among those ridiculing the theory of distributed memory was Indiana University biologist Paul Pietsch. In earlier experiments, Pietsch had discovered that he could remove the brain of a salamander and although the animal became comatose, it would resume functioning once the brain was put back in. If Pribram were right, then some of the salamander's brain could be removed, or reshuffled, and it shouldn't affect its ordinary function. But Pietsch was certain that Pribram was wrong and he was fierce in his determination to prove it so. In more than 700 experiments, Pietsch cut out scores of salamander brains. Before putting them back in, he began tampering with them. In successive experiments he reversed, cut out, sliced away, shuffled and even sausage-ground his test subjects' brains. But no matter how brutally mangled, or diminished in size, whenever whatever was left of the brains were returned to his subjects and the salamanders had recovered, they returned to normal behavior. From being a complete skeptic, Pietsch turned convert to Pribram's view that memory is distributed throughout the brain.[20]

Pribram's theories were also vindicated in 1979 by a husband-and-wife team of neurophysiologists at the University of California at Berkeley. Russell and Karen DeValois converted simple plaid and checkerboard patterns into Fourier waves and discovered that the brain cells of cats and monkeys responded not to the patterns themselves but to the interference patterns of their component waves. Countless studies, elaborated on by the DeValois team in their book *Spatial Vision*,[21] show that numerous cells in the visual system are tuned into certain frequencies. Other studies by Fergus Campbell of Cambridge University in England, as well as by a number of other laboratories, also showed that the cerebral cortex of humans may be tuned to specific frequencies.[22] This would explain how we can recognize things as being the same, even when they are vastly different sizes.

Pribram also showed that the brain is a highly discriminating frequency analyzer. He demonstrated that the brain contains a certain 'envelope', or mechanism, which limits the otherwise infinite wave information available to it, so that we are not bombarded with limitless wave information contained in the Zero Point Field.[23]

In his own studies in the laboratory, Pribram confirmed that the visual cortex of cats and monkeys responded to a limited range of frequencies.[24] Russell DeValois and his colleagues also showed that the receptive fields in the neurons of the cortex were tuned to a very small range of frequencies.[25] In his studies of both cats and humans, Campbell at Cambridge also demonstrated that neurons in the brain responded to a limited band of frequencies.[26] At one point, Pribram came across the work of the Russian Nikolai Bernstein. Bernstein had made films of human subjects dressed entirely in black costumes on which white tapes and dots had been placed to mark the limbs – not unlike the classic Halloween skeleton costume. The participants were asked to dance against a black background while being filmed. When the film was processed, all that could be seen was a series of white dots moving in a continuous pattern in a wave form. Bernstein analyzed the waves. To his astonishment, all the rhythmic movements could be represented in Fourier trigonometric sums to such an extent that he found that he could predict the next movements of his dancers 'to an accuracy of within a few millimeters'.[27]

The fact that movement could somehow be represented formally in terms of Fourier equations made Pribram realize that the brain's conversations with the body might also be occurring in the form of waves and patterns, rather than as images.[28] The brain somehow had the capacity to analyze movement, break it down into wave frequencies and transmit this wave-pattern shorthand to the rest of the body. This information, transmitted nonlocally, to many parts at once, would explain how we can fairly easily manage complicated global tasks involving multiple body parts, such as riding a bicycle or roller skating. It also accounts for how we can easily imitate some task. Pribram also came across evidence that our other senses – smell, taste and hearing – operate by analyzing frequencies.[29]

In Pribram's own studies with cats, in which he recorded frequencies from the motor cortex of cats while their right forepaw was being moved and up down, he discovered that, like the visual cortex, individual cells in the cat's motor cortex responded to only a limited number of frequencies of movement, just as individual strings in a piano respond to a limited range of frequencies.[30]

Pribram struggled with where this intricate process of wavefront decoding and transformation could possibly take place. It then occurred to him that the one area of the brain where wave-interference patterns might be created was not in any particular cell, but in the spaces between them. At the end of every neuron, the basic unit of a brain cell, are synapses, where chemical charges build up, eventually triggering electrical firing across these spaces to the other neurons. In the same spaces, dendrites – tiny filaments of nerve endings wafting back and forth, like shafts of wheat in a slow breeze – communicate with other neurons, sending out and receiving their own electrical wave impulses. These 'slow-wave potentials', as they are called, flow through the glia, or glue, surrounding neurons, to gently touch or even collide with other waves. It is at this busy juncture, a place of a ceaseless scramble of electromagnetic communications between synapses and dendrites, where it was most likely that wave frequencies could be picked up and analyzed, and holographic images formed, since these wave patterns criss-crossing all the time are creating hundreds and thousands of wave-interference patterns.

Pribram conjectured that these wave collisions must create the pictorial images in our brain. When we perceive something, it's not due to the activity of neurons themselves but to certain patches of dendrites distributed around the brain, which, like a radio station, are set to resonate only at certain frequencies. It is like having a vast number of piano strings all over your head, only some of which would vibrate as a particular note is played.

Pribram largely left it to others to test his views so that he wouldn't jeopardize his more traditional laboratory work by being associated with his own revolutionary notions. For some years his theory languished. He would have to wait several decades after his initial proposal for other pioneers in the scientific community to catch up with him. His most important support was from an unlikely source: a German trying to make a medical diagnostic machine work better.

Walter Schempp, a mathematics professor from the University of Siegen in Germany, believed he was simply carrying on the work of his ancestor Johannes Kepler, an astronomer working in the sixteenth and seventeenth centuries. Kepler famously claimed in his book *Harmonice mundi*, that people on earth could hear the music of the stars. At the time, Kepler's contemporaries thought him crazy. It was four hundred years before a pair of American scientists showed that there is indeed a music of the heavens. In 1993, Hulse and Taylor landed the Nobel prize for discovering binary pulsars – stars which send out electromagnetic waves in pulses. The most sensitive of equipment located in one of the world's highest places, high on a mountaintop in Arecibo, Puerto Rico, picks up evidence of their existence through radio waves.

As a nod to his forebear, Walter himself had specialized in the mathematics of harmonic analysis, or the frequency and phase of sound waves. It occurred to him one day, sitting at home in his garden – his three-year-old son was ill at the time – that you might be able to extract three-dimensional images from sound waves. Without reading of Gabor, he'd worked out his own holographic theory, reconstructed from mathematical theory. He'd consulted his own books in mathematics to no avail, but after looking up what had been done in optical theory, he came across Gabor's work.

By 1986, Walter had published a book which proved mathematically how you could get a hologram from the echoes of the radio waves received in radar, which came to be regarded as a classic in state-of-the art radar. Schempp began thinking that the same principles of wave holography might apply to magnetic resonance imaging (MRI), a medical tool used to examine the soft tissues of the body, which was still in its infancy. But when he inquired about it, he soon realized that the people who'd developed and were running the machines had little idea how MRI worked. The technology was so primitive that it was simply being used intuitively. Patients would have to sit still for four hours or more while pictures were slowly taken, by what means nobody was exactly sure. Walter was utterly dissatisfied with MRI technology as it then stood and realized that it was a relatively simple prospect to make sharper images.

To do so, however, required an incredible commitment from the then 50-year-old, who, despite having a young family, with his greying hair and melancholic nature already looked more mature than his years. He had to study medicine, biology and radiology in order to become trained as a doctor before being able to use the equipment. He accepted a place offered at Johns Hopkins Medical School in Baltimore, Maryland, which has the best outpatient radiology department in the USA, and later trained at Massachusetts General Hospital, which is affiliated with MIT. After a fellowship in radiology in Zurich, Walter was finally able to return to Germany, where he now had the appropriate qualifications to officially lay hands on the machine.

Taking pictures of the brain and soft tissues of the body with MRI is ordinarily a matter of getting to the water lurking in the various nooks and crevices. To do so, you need to be able to find the nuclei of the water molecules scattered throughout the brain. Because protons spin, like little magnets, locating them is often most simply accomplished by applying a magnetic field. This causes the spin to accelerate, eventually to the point where the nuclei behave like microscopic gyroscopes spinning out of control. All this molecular manipulation makes the water molecules that much more conspicuous, enabling the MRI machine to locate them and ultimately to extract an image of the brain's soft tissues.

As the molecules slow down, they give off radiation. What Walter discovered is that this radiation contained encoded wave information about the body, which the machine can capture and eventually use to reconstruct a three-dimensional image of the body. The information that you extract is an encoded hologram of a slice of the brain or body part that you wish to examine. Through the use of Fourier transforms, and many slices of the body, you combine and eventually turn this information into an optical picture.

Schempp went on to help revolutionize the construction of MRI machines and wrote a textbook on the subject, showing that imaging worked as holography did, and he would soon become the world authority on the machine and functional MRI, which allows you to actually observe brain activity elicited by sensory stimuli.[31] His improvements cut down the time required for a patient to sit still from 4 hours to 20 minutes. But he began to wonder whether the mathematics and theory of how this machine worked could be applied to biological systems. He had called his theory 'quantum holography', because what he'd really discovered was that all sorts of information about objects, including their three-dimensional shape, is carried in the quantum fluctuations of the Zero Point Field, and that this information can be recovered and reassembled into a three-dimensional image. Schempp had discovered, as Puthoff had predicted, that the Zero Point Field was a vast memory store. Through Fourier transformation, MRI machines could take information encoded in the Zero Point Field and turn it into images. The real question he was posing went far beyond whether he could create a sharper image in MRI. What he was really trying to find out was whether his mathematical equations unlocked to the key to the human brain.

In his quest to apply his theories to something larger, Walter came across the work of Peter Marcer, a British physicist who'd worked as a student and colleague of Dennis Gabor and gone on to CERN in

Switzerland. Marcer himself had been doing some work on a computation based on wave theory in sound, and he was sitting there with a theory, which he intuitively sensed could be applied to the human brain. The problem was that the theory was abstract and general, and needed more mathematical rooting to make it concrete. In the early 1990s, he received a call from Walter Schempp, whose work threw a life jacket to his theory. It grounded his own work into something tidy and mathematical.

In Marcer's mind, Walter's machine worked on the same principle that Karl Pribram had worked out for the human brain: by reading natural radiation and emissions from the Zero Point Field. Not only did Walter have a mathematical map of how information processing in the brain may work, which amounted to a mathematical demonstration of the theories of Karl Pribram. He also had, as Peter saw it, a machine which worked according to this process. Like Pribram's model of the brain, Schempp's MRI machine underwent a staged process, combining wave-interference information taken from different views of the body and then eventually transforming it into a virtual image. MRI was experimental verification that Peter's own quantum mechanical theory actually worked.

Although Walter had written some general papers about how his work could be applied to biological systems, it was only in partnership with Peter that he began to apply his theory to a theory of nature and the individual cell. They wrote papers together, each time refining their theories. Two years later, Peter was at a conference and heard Edgar Mitchell speak about his own theory of nature and human perception, which sounded serendipitously similar to his own. They spent several excited lunches comparing notes and decided that all three of them needed to collaborate. Walter would also correspond with Pribram, trading information. What they all discovered was something that Pribram's work had always hinted at: perception occurred at a much more fundamental level of matter – the netherworld of the quantum particle. We didn't see objects *per se*, but only their quantum information and out of that constructed our image of the world. Perceiving the world was a matter of tuning into the Zero Point Field.

Stuart Hameroff, an anesthesiologist from the University of Arizona, had been thinking about how anesthetic gases turn off consciousness. It fascinated him that gases with such disparate chemistry as nitrous oxide

(N_2O), ether ($CH_3CH_2OCH_2CH_3$), halothane ($CF_3CHClBr$), chloroform ($CHCl_3$) and isoflurane ($CHF_2OCHClCF_3$) could all bring about loss of consciousness.[32] It must have something to do with some property besides chemistry. Hameroff guessed that general anesthetics must interfere with the electrical activity within the microtubules, and this activity would turn off consciousness. If this were the case, then the reverse would also be true: electrical activity of microtubules that composed the insides of dendrites and neurons in the brain must somehow be at the heart of consciousness.

Microtubules are the scaffolding of the cell, maintaining its structure and shape. These microscopic hexagonal lattices of fine filaments of protein, called tubulins, form tiny hollow cylinders of indefinite length. Thirteen strands of tubules wrap around the hollow core in a spiral; and all the microtubules in a cell radiate outward from the center to the cell membrane, like a cartwheel. We know that these little honeycomb structures act as tracks in transporting various products along cells, particularly in nerve cells, and they are vital for pulling apart chromosomes during cell division. We also know that most microtubules are constantly remaking themselves, assembling and disassembling, like an endless set of Lego.

In his own experiments with the brains of small mammals, Hameroff found, like Fritz Popp, that living tissue was transmitting photons and that good penetration of 'light' occurred in certain areas of the brain.[33]

Microtubules appeared to be exceptional conductors of pulses. Pulses sent in one end traveled through pockets of protein and arrived unchanged at the other. Hameroff also discovered a great degree of coherence among neighboring tubules, so that a vibration in one microtubule would tend to resonate in unison through its neighbors.

It occurred to Hameroff that the microtubules within the cells of dendrites and neurons might be 'light pipes', acting as 'waveguides' for photons, sending these waves from cell to cell throughout the brain without any loss of energy. They might even act as tiny tracks for these light waves throughout the body.[34]

By the time that Hameroff began formulating his theory, many of Pribram's ideas, which had been so outrageous when he had first formulated them, were being taken up in many quarters. Scientists in research centers around the globe were beginning to concur that the brain made use of quantum processes. Kunio Yasue, a quantum physicist from Kyoto, Japan, had carried out mathematical formulations to help understand the neural microprocess. Like Pribram, his equations showed that brain processes occurred at the quantum level, and that the dendritic networks in the brain were operating in tandem through quantum coherence. The equations developed in quantum physics precisely described this cooperative interaction.[35] Independently of Hameroff, Yasue and his colleague Mari Jibu, of the Department of Anesthesiology, Okayama University, in Japan, had also theorized that the quantum messaging of the brain must take place through vibrational fields, along the microtubules of cells.[36] Others had theorized that the basis of all the brain's functions had to do with the interaction between brain physiology and the Zero Point Field.[37] An Italian physicist, Ezio Insinna of the Bioelectronics Research Association, in his own experimental work with microtubules, discovered that these structures had a signaling mechanism, thought to be associated with the transfer of electrons.[38]

Eventually, many of these scientists, each of whom seemed to have one piece of the puzzle, decided to collaborate. Pribram, Yasue, Hameroff and Scott Hagan from the Department of Physics at McGill University assembled a collective theory about the nature of human consciousness.[39] According to their theory, microtubules and the membranes of dendrites represented the Internet of the body. Every neuron of the brain could log on at the same time and speak to every other neuron simultaneously via the quantum processes within.

Microtubules helped to marshal discordant energy and create global coherence of the waves in the body – a process called 'superradiance' – then allowed these coherent signals to pulse through the rest of the body. Once coherence was achieved, the photons could travel all along the light pipes as if they were transparent, a phenomenon called 'self-induced transparency'. Photons can penetrate the core of the microtubule and communicate with other photons throughout the body, causing collective cooperation of subatomic particles in microtubules throughout the brain. If this is the case, it would account for unity of thought and consciousness – the fact that we don't think of loads of disparate things at once.[40]

Through this mechanism, the coherence becomes contagious, moving from individual cells to cell assemblies – and in the brain from certain neuron cell assemblies to others. This would provide an explanation for the instantaneous operation of our brains, which occurs at between one ten-thousandth and one-thousandth of a second, requiring that information be transmitted at 100–1000 metres per second – a speed that exceeds the capabilities of any known connections between axons or dendrites in neurons. Superradiance along the light pipes also could account for a phenomenon that has long been observed – the tendency of EEG patterns in the brain to get synchronized.[41]

Hameroff observed that electrons glide easily along these light pipes without getting entangled in their environment – that is, settling into any set single state. This means they can remain in a quantum state – a condition of all possible states – enabling the brain eventually to finally choose among them. This might be a good explanation for free will. At every moment, our brains are making quantum choices, taking potential states and making them actual ones.[42]

It was only a theory – it hadn't undergone the exhaustive testing of Popp and his biophoton emissions – but some good mathematics and circumstantial evidence gave it weight. The Italian physicists Del Giudice and Preparata had also come up with some experimental evidence of Hameroff's theory that light pipes contained coherent energy fields inside them.

Microtubules are hollow and empty save for some water. Ordinary water, from a tap or in a river, is disordered, with molecules that move randomly. But some of the water molecules in brain cells are coherent, the Italian team discovered, and this coherence extends as far as 3 nanometres or more outside the cell's cytoskeleton. Since this is the case, it is overwhelmingly likely that the water inside the microtubules is also ordered. This offered indirect evidence that some sort of quantum process, creating quantum coherence, was occurring inside.[43] They'd also shown that this focusing of waves would produce beams 15 nanometres in diameter – precisely the size of the microtubule's inner core.[44]

All of this led to a heretical thought, which had already occurred to Fritz-Albert Popp. Consciousness was a global phenomenon that occurred everywhere in the body, and not simply in our brains. Consciousness, at its most basic, was coherent light.

Although each of the scientists – Puthoff, Popp, Benveniste and Pribram – had been working independently, Edgar Mitchell was one of the few to realize that, as a totality, their work presented itself as a unified theory of mind and matter – evidence of physicist David Bohm's vision of a world of 'unbroken wholeness'.[45] The universe was a vast dynamic cobweb of energy exchange, with a basic substructure containing all possible versions of all possible forms of matter. Nature was not blind and mechanistic, but open-ended, intelligent and purposeful, making use of a cohesive learning feedback process of information being fed back and forth between organisms and their environment. Its unifying mechanism was not a fortunate mistake but information which had been encoded and transmitted everywhere at once.[46]

Biology was a quantum process. All the processes in the body, including cell communication, were triggered by quantum fluctuations, and all higher brain functions and consciousness also appeared to function at the quantum level. Walter Schempp's explosive discovery about quantum memory set off the most outrageous idea of all: short- and long-term memory doesn't reside in our brain at all, but instead is stored in the Zero Point Field. After Pribram's discoveries, a number of scientists, including systems theorist Ervin Laszlo, would go on to argue that the brain is simply the retrieval and read-out mechanism of the ultimate storage medium – The Field.[47] Pribram's associates from Japan would hypothesize that what we think of as memory is simply a coherent emission of signals from the Zero Point Field, and that longer memories are a structured grouping of this wave information.[48] If this were true, it would explain why one tiny association often triggers a riot of sights, sounds and smells. It would also explain why, with long-term memory in particular, recall is instantaneous and doesn't require any scanning mechanism to sift though years and years of memory.

If they are correct, our brain is not a storage medium but a receiving mechanism in every sense, and memory is simply a distant cousin of ordinary perception. The brain retrieves 'old' information the same way it processes 'new' information – through holographic transformation of wave interference patterns.[49] Lashley's rats with the fried brains were able to conjure up their run in its entirety because the memory of it was never burned away in the first place. Whatever reception mechanism was left in the brain – and as Pribram had demonstrated, it was distributed all over the brain – was tuning back into the memory through The Field.

Some scientists went as far as to suggest that all of our higher cognitive processes result from an interaction with the Zero Point Field.[50] This kind of constant interaction might account for intuition or creativity – and how ideas come to us in bursts of insight, sometimes in fragments but often as a miraculous whole. An intuitive leap might simply be a sudden coalescence of coherence in The Field.

The fact that the human body was exchanging information with a mutable field of quantum fluctuation suggested something profound about the world. It hinted at human capabilities for knowledge and communication far deeper and more extended than we presently understand. It also blurred the boundary lines of our individuality – our very sense of separateness. If living things boil down to charged particles interacting with a field and sending out and receiving quantum information, where did we end and the rest of the world begin? Where was consciousness – encased inside our bodies or out there in The Field? Indeed, there was no more 'out there' if we and the rest of the world were so intrinsically interconnected.

The implications of this were too huge to ignore. The idea of a system of exchanged and patterned energy and its memory and recall in the Zero Point Field hinted at all manner of possibility for human beings and their relation to their world. Modern physicists had set mankind back for many decades. In ignoring the effect of the Zero Point Field, they'd eliminated the possibility of interconnectedness and obscured a scientific explanation for many kinds of miracles. What they'd being doing, in renormalizing their equations, was a little like subtracting out God.

Part 2

The Extended Mind

You are the world.

Krishnamurti

CHAPTER SIX

The Creative Observer

IT IS STRANGE WHAT CLINGS to your mind from the flotsam and jetsam of the everyday. For Helmut Schmidt it was an article in, of all places, *Reader's Digest*. He'd read it as a 20-year-old student in 1948, at the University of Cologne, after Germany had just emerged from the Second World War. It lodged in his memory for nearly twenty years, surviving through two emigrations, from Germany to America and from academia to industry – from a professorship at the University of Cologne to a position as a research physicist at Boeing Scientific Research Laboratories in Seattle, Washington.

Through all his changes of country and career, Schmidt pondered the meaning of the article, as though something in him knew that it was central to his life's direction even before he was consciously aware of it. Every so often he would engage in a bit more reflection, take out the article in his mind's eye and examine it in the light, turning it this way and that, before filing it away again, a bit of unfinished business he wasn't yet sure how to tend to.[1]

The article had been nothing more than an abridged version of some writing by the biologist and parapsychologist J. B. Rhine. It concerned his famous experiments on precognition and extrasensory perception, including the card tests which would later be used by Edgar Mitchell in outer space. Rhine had conducted all of his experiments under carefully controlled conditions and they had yielded interesting results.[2] The studies had shown that it was possible for a person to transmit information about card symbols to another or increase the odds of a certain number being rolled with a set of dice.

Schmidt had been drawn to Rhine's work for its implications in physics. Even as a student, Schmidt had had a contrary streak, which rather liked testing the limits of science. In his private moments, he regarded physics and many of the sciences, with their claim to have explained many of the mysteries of the universe, as exceedingly presumptuous. He'd been most interested in quantum physics, but he found himself perversely drawn to those aspects of quantum theory which presented the most potential problems.

What held the most fascination of all for Schmidt was the role of the observer.[3] One of the most mysterious aspects of quantum physics is the so-called Copenhagen interpretation (so named because Niels Bohr, one of the founding fathers of quantum physics, resided there). Bohr, who forcefully pushed through a variety of interpretations in quantum physics without the benefit of a unified underlying theory, set out various dictums about the behavior of electrons as a result of the mathematical equations which are now followed by workaday physicists all over the world. Bohr (and Werner Heisenberg) noted that, according to experiment, an electron is not a precise entity, but exists as a potential, a superposition, or sum, of all probabilities until we observe or measure it, at which point the electron freezes into a particular state. Once we are through looking or measuring, the electron dissolves back into the ether of all possibilities.

Part of this interpretation is the notion of 'complementarity' – that you can never know everything about a quantum entity such as an electron at the same time. The classic example is position and velocity; if you discover information about one aspect of it – where it is, for instance – you cannot also determine exactly where it's going or at what speed.

Many of the architects of quantum theory had grappled with the larger meaning of the results of their calculations and experiments, making comparisons with metaphysical and Eastern philosophical texts.[4] But the rank and file of physicists in their wake complained that the laws of the quantum world, while undoubtedly correct from a mathematical point of view, beggared ordinary common sense. French physicist and Nobel prize winner Louis de Broglie devised an ingenious thought experiment, which carried quantum theory to its logical conclusion. On the basis of current quantum theory, you could place an electron in a container in Paris, divide the container in half, ship one half to Tokyo and the other to New York, and, theoretically, the electron should still occupy either side unless you peer inside, at which point a definite position in one half or the other would finally be determined.[5]

What the Copenhagen interpretation suggested was that randomness is a basic feature of nature. Physicists believe this is demonstrated by another famous experiment involving light falling on a semi-transparent mirror. When light falls on such a mirror, half of it is reflected and the other half is transmitted through it. However, when a single photon arrives at the mirror, it must go one way or the other, but the way it will go – reflected or transmitted – cannot be predicted. As with any such binary process, we have a 50–50 chance of guessing the eventual route of the photon.[6] On the subatomic level, there is no causal mechanism in the universe.

If that were so, Schmidt wondered, how was it that some of Rhine's subjects were able to correctly guess cards and dice – implements, like a photon, of random processes? If Rhine's studies were correct, something fundamental about quantum physics was wrong. So-called random binary processes could be predicted, even influenced.

What appeared to put a halt to randomness was a living observer. One of the fundamental laws of quantum physics says that an event in the subatomic world exists in all possible states until the act of observing or measuring it 'freezes' it, or pins it down, to a single state. This process is technically known as the collapse of the wave function, where 'wave function' means the state of all possibilities. In Schmidt's mind, and the minds of many others, this was where quantum theory, for all its mathematical perfection, fell down. Although nothing existed in a single state independently of an observer, you could describe what the observer sees, but not the observer himself. You included the moment of observation in the mathematics, but not the consciousness doing the observing. There was no equation for an observer.[7]

There was also the ephemeral nature of it all. Physicists couldn't offer any real information about any given quantum particle. All they could say with certainty was that when you took a certain measurement at a certain point, this is what you would find. It was like catching a butterfly on the wing. Classical physics didn't have to talk about an observer; according to Newton's version of reality, a chair or even a planet was sitting there, whether or not we were looking at it. The world existed out there independently of us.

But in the strange twilight of the quantum world, you could only determine incomplete aspects of subatomic reality with an observer pinning down a single facet of the nature of an electron only at that moment of observation, not for all time. According to the mathematics, the quantum world was a perfect hermetic world of pure potential, only made real – and, in a sense, less perfect – when interrupted by an intruder.

It seems to be a truism of important shifts in thinking that many minds begin to ask the same question at roughly the same time. In the early 1960s, nearly twenty years after he'd first read Rhine's article, Schmidt, like Edgar Mitchell, Karl Pribram and the others, was one of a growing number of scientists trying to get some measure of the nature of human consciousness in the wake of the questions posed by quantum physics and the observer effect. If the human observer settled an electron into a set state, to what extent did he or she influence reality on a large scale? The observer effect suggested that reality only emerged from a primordial soup like the Zero Point Field with the involvement of living consciousness. The logical conclusion was that the physical world only existed in its concrete state while we were involved in it. Indeed, Schmidt wondered, was it true that nothing existed independently of our perception of it?

A few years after Schmidt was pondering all this, Mitchell would head off to Stanford on the West Coast of the USA, gathering funding for his own consciousness experiments with a number of gifted psychics. For Mitchell, like Schmidt, the importance of Rhine's findings would be what they appeared to show about the nature of reality. Both scientists wondered to what extent order in the universe was related to the actions and intentions of human beings.

If consciousness itself created order – or indeed in some way created the world – this suggested much more capacity in the human being than was currently understood. It also suggested some revolutionary notions about humans in relation to their world and the relation between all living things. What Schmidt was also asking was how far our bodies extended. Did they end with what we always thought of as our own isolated persona, or 'extend out' so that the demarcation between us and our world was less clear-cut? Did living consciousness possess some quantum-field-like properties, enabling it to extend its influence out into the world? If so, was it possible to do more than simply observe? How strong was our influence? It was only a small step in logic to conclude that in our act of participation as an observer in the quantum world, we might also be an influencer, a creator.[8] Did we not only stop the butterfly at a certain point in its flight, but also influence the path it will take – nudging it in a particular direction?

A related quantum effect suggested by Rhine's work was the possibility of nonlocality, or action at a distance: the theory that two subatomic particles once in close proximity seemingly communicate over any distance after they are separated. If Rhine's ESP experiments were to be believed, action at a distance might also be present in the world at large.

Schmidt was 37 before he finally got the opportunity to test out his ideas, in 1965, during his tenure at Boeing. A tall, thin presence with a pronounced, angular intensity, his hair heavily receded on either side of an exaggerated widow's peak, Schmidt was in the happy circumstance of being employed to pursue pure research in the Boeing laboratory, whether or not it was connected to aerospace development. Boeing was in a lull in its fortunes. The aerospace giant had come up with the supersonic but had shelved it, and hadn't yet invented the 747, so Schmidt had time on his hands.

An idea slowly began taking shape. The simplest way to test all these ideas was to see if human consciousness could affect some sort of probabilistic system, as Rhine had done. Rhine had used his special cards for the ESP 'forced choice' guessing, or 'precognition', exercises and dice for 'psychokinesis' – tests of whether mind could influence matter. There were certain limitations with both media. You could never truly show that a toss of the dice had been a random process affected by human consciousness, or that a correct guess of the face of a card hadn't been purely down to chance. Cards might not be shuffled perfectly, a die might be shaped or weighted to favor a certain number. The other problem was that Rhine had recorded the results by hand, a process that could be prone to human error. And finally, because they were done manually, the experiments took a long time.

Schmidt believed he could contribute to Rhine's work by mechanizing the testing process. Because he was considering a quantum effect, it made sense to build a machine whose randomness would be determined by a quantum process. Schmidt had read about two Frenchmen, named Remy Chauvin and Jean-Pierre Genthon, who'd conducted studies to see if their test subjects could in some way change the decay rate of radioactive materials, which would be recorded by a Geiger counter.[9]

Nothing much is more random than radioactive atomic decay. One of the axioms of quantum physics is that no one can predict exactly when an atom will decay and an electron consequently be released. If Schmidt made use of radioactive decay in the machine's design, he could produce what was almost a contradiction in terms: a precision instrument built upon quantum mechanical uncertainty.

With machines using a quantum decay process, you're dealing in the realm of probability and fluidity – a machine governed by atomic particles, in turn governed by the probabilistic universe of quantum mechanics. This would be a machine whose output consisted of perfectly random activity, which in physics is viewed as a state of 'disorder'. The Rhine studies in which participants had apparently affected the roll of the dice suggested that some information transfer or ordering mechanism was going on – what physicists like to term 'negative entropy', or 'negentropy' for short – the move away from randomness, or disarray, to order. If it could be shown that participants in a study had altered some element of the machine's output, they'd have changed the probabilities of events – that is, shifted the odds of something happening or altered the tendency of a system to behave in a certain way.[10] It was like persuading a person at a crossroads, momentarily undecided about taking a walk, to head down one road rather than another. They would, in other words, have created order.

As most of his work had consisted of theoretical physics, Schmidt needed to brush up on his electronics in order to construct his machine. With the help of a technician, he produced a small, rectangular box, slightly larger than a fat hardback book, with four colored lights and buttons and a thick cable attached to another machine punching coding holes in a stream of paper tape. Schmidt dubbed the machine a 'random number generator', which he came to refer to as RNG. The RNG had the four colored lights on top of it – red, yellow, green and blue – which would flash on randomly.

In the experiment, a participant would press a button under one of the lights, which registered a prediction that the light above it would light up.[11] If you were correct, you'd score a hit. On top of the device were two counters. One would count the number of 'hits' – the times the participant could correctly guess which lamp would light – and the other would count the number of trials. Your success rate would be staring at you as you continued with the experiment.

Schmidt had employed a small amount of the isotope strontium-90, placed near an electron counter so that any electrons ejected from the unstable, decaying molecules would be registered inside a Geiger–Müller tube. At the point where an electron was flung into the tube – at a rate, on average, of 10 a second – it stopped a high-speed counter breathlessly racing through numbers between one and four at a million per second, and the number stopped at would light the correspondingly numbered lamp. If his participants were successful, it meant that they had somehow intuited the arrival time of the next electron, resulting in the lighting of their designated lamp.

If someone was just guessing, he'd have a 25 per cent chance of getting the right results. Most of Schmidt's first test subjects scored no better than this, until he contacted a group of professional psychics in

Seattle and collected subjects who went on to be successful. Thereafter, Schmidt was meticulous in his recruitment of participants with an apparent psychic gift for guessing correctly. The effects were likely to be so minuscule, he figured, that he had to maximize his chances of success. With his first set of studies, Schmidt got 27 per cent – a result that may appear insignificant, but which was enough of a deviation in statistical terms for him to conclude that something interesting was going on.[12]

Apparently, there'd been some connection between the mind of his subjects and his machine. But what was it? Did his participants foresee which lights would be lit? Or did they make a choice among the colored lamps and somehow mentally 'force' that particular lamp to light? Was the effect precognition or psychokinesis?

Schmidt decided to isolate these effects further by testing psychokinesis. What he had in mind was an electronic version of Rhine's dice studies. He went on to build another type of machine – a twentieth-century version of the flip of a coin. This machine was based on a binary system (a system with two choices: yes or no; on or off; one or zero). It could electronically generate a random sequence of 'heads' and 'tails' which were displayed by the movement of a light in a circle of nine lamps. One light was always lit. With the top lamp lit at the start, for each generated head or tail the light moved by one step in a clockwise or anticlockwise direction. If 'heads' were tossed, the next light in clockwise order would light. If 'tails', the next light in the anticlockwise direction would light instead. Left to its own devices, the machine would take a random walk around the circle of nine lights, with movements in each direction roughly half the time. After about two minutes and 128 moves, the run stopped and the numbers of generated heads and tails were displayed. The full sequence of moves was also recorded automatically on paper tape, with the number of heads or tails indicated by counters.

Schmidt's idea was to have his participants will the lights to take more steps in a clockwise direction. What he was asking his participants to do, on the most elementary level, was to get the machine to produce more heads than tails.

In one study, Schmidt worked with two participants, an aggressive, extroverted North American woman and a reserved male researcher in parapsychology from South America. In preliminary tests, the North American woman had scored consistently more heads than tails, while the South American man had scored the reverse – more tails than heads –

even though he'd been trying for a greater number of heads. During a larger test of more than 100 runs apiece, both kept to the same scoring tendencies – the woman got more heads, the man more tails. When the woman did her test, the light showed a preference for clockwise motion 52.5 per cent of the time. But when the man concentrated, the machine once again did the opposite of what he intended. In the end, only 47.75 per cent of the lit lights moved in a clockwise direction.

Schmidt knew he had come up with something important, even if he couldn't yet put his finger on how any known law of physics could explain this. When he worked it out, the odds against such a large disparity in the two scores occurring by chance were more than 10 million to one. That meant he'd have to conduct 10 million similar studies before he'd get the results by chance alone.[13]

Schmidt gathered together eighteen people, the most easily available he could find. In their first studies, he found that, as with his South American fellow, they seemed to have a reverse effect on the machine. If they tried to make the machine move clockwise, it tended to move in the other direction.

Schmidt was mainly interested in whether there was any effect at all, no matter what the direction. He decided to see whether he could set up an experiment to make it more likely that his subjects got a negative score. If these participants ordinarily had a negative effect, then he'd do his best to amplify it. He selected only those participants who'd had a reverse effect on the machine. He then created an experimental atmosphere that might encourage failure. His participants were asked to conduct their test in a small dark closet where they'd be huddled with the display panel. Schmidt studiously avoided giving them the slightest bit of encouragement. He even told them to expect that they were going to fail.

Not surprisingly, the team had a significantly negative effect on the RNG. The machine moved more in the opposite way than what they'd intended. But the point was that the participants were having some effect on the machine, even if it was a contrary one. Somehow, they'd been able to shift the machines, ever so slightly, away from their random activity; their results were 49.1 per cent against an expected result of 50 per cent. In statistical terms, this was a result of major significance – a thousand to one that the result had occurred by chance. Since none of his subjects knew how the RNG worked, it was clear that whatever they were doing must have been generated by some sort of human will.[14]

Schmidt carried on with similar studies for a number of years, publishing in *New Scientist* and other journals, meeting with like-minded people and achieving highly significant scores in his studies – sometimes as high as 54 per cent against an expected result of 50 per cent.[15] By 1970, the year before Mitchell's moon walk, Boeing suffered a setback in profits and needed to cut back sharply on staff. Schmidt, along with hundreds of others, was one of its casualties. Boeing had been such a key source of R&D jobs in the area that without the aerospace giant, there was virtually no work to be had. A sign at the border of Seattle read, 'Will the last one to leave Seattle please turn off the lights?' Schmidt made his third and final career move. He would continue on with his consciousness research, a physicist among parapsychologists. He relocated to Durham, North Carolina, and sought work at Rhine's laboratory, the Foundation for Research on the Nature of Man, carrying on his RNG research with Rhine himself.

A few years later, word of Schmidt's machines filtered through to Princeton University and came to the attention of a young university student in the school of engineering. She was an undergraduate, a sophomore, studying electrical engineering, and something about the idea of mind being able to influence a machine held a certain romantic appeal. In 1976, she decided to approach the dean of the engineering school about the possibility of replicating Helmut Schmidt's RNG studies as a special project.[16]

Robert Jahn was a tolerant man. When campus unrest had erupted at Princeton, as it did at most universities across America in response to the escalation of the Vietnam War, Jahn, then a professor of engineering, had found himself an unwitting apologist for high technology, at a point when it was being blamed for America's stark polarization. Jahn had argued persuasively to the Princeton student body that technology actually offered the solution to this divisiveness. His conciliatory line not only had settled down the campus unrest but also had helped to create an accepting atmosphere for students with technical interests at what was essentially a liberal arts university. Jahn's skill at diplomacy may have been one reason he'd been asked to serve as dean in 1971.

Now his famous tolerance was being stretched nearly to its limit. Jahn was an applied physicist who had invested his entire life in the teaching and development of technology. All of his own degrees came from Princeton, and his work in advanced space propulsion systems and high temperature plasma dynamics had won him his current distinguished position.

He'd returned to Princeton in the early 1960s with the mission of introducing electric propulsion to the aeronautical engineering department. The project he was now being asked to supervise essentially belonged to the category of psychic phenomena. Jahn wasn't convinced it was a viable topic, but the sophomore was such a brilliant student who was already on a fast track through her program that he eventually relented. He agreed to subsidize a summer project for her out of his discretionary funds. Her task was to research the existing scientific literature on RNG studies and other forms of psychokinesis and to carry out a few preliminary experiments. If she could convince Jahn that the field held some credibility and, more importantly, could be approached from a technical perspective, he told her, then he'd agree to supervise her independent work.

Jahn tried to approach the topic as an open-minded scholar might. Over the summer, his student would leave photocopies of technical papers on his desk and even managed to coax him into accompanying her to a meeting of the Parapsychological Association. He tried to get a feel for the people involved in studying what had always been dismissed as a fringe science. Jahn rather hoped that the entire subject would go away. Much as he was amused by the project, particularly by the notion that he somehow might have the power to influence all the complicated array of equipment around him, he knew that this was something, in the long run, that might mean trouble for him, particularly among his fellow faculty members. How would he ever explain it as a serious topic of study?

Jahn's student kept returning with more convincing proof that this phenomenon existed. There was no doubt that the people involved in the studies and the research itself had a certain credibility. He agreed to supervise a two-year project for her, and when she began returning with her own successful results, he found himself making suggestions and trying to refine the equipment.

By the second year of the student's project, Jahn himself began dabbling in his own RNG experiments. It was beginning to look as though there might be something interesting here. The student graduated and left her RNG work behind, an intriguing thought experiment, and no more, the results of which had satisfied her curiosity. Now it was time to get serious and return to the more traditional line she'd originally chosen for herself. She embarked on what would turn out to be a lucrative career in conventional computer science, leaving in her wake a body of tantalizing data and also a bomb across Bob Jahn's path that would change the course of his life forever.

Jahn respected many of the investigators into consciousness research, but privately he felt that they were going about it the wrong way. Work like Rhine's, no matter how scientific, tended to be placed under the general umbrella of parapsychology, which was largely dismissed by the scientific establishment as the province of confidence tricksters and magicians. Clearly what was needed was a highly sophisticated, solidly based research program, which would give the studies a more temperate and scholarly framework. Jahn, like Schmidt, realized the enormous implications of these experiments. Ever since Descartes had postulated that mind was isolated and distinct from the body, all the various disciplines of science had made a clear distinction between mind and matter. The experiments with Schmidt's machines seemed to be suggesting that this separation simply didn't exist. The work that Jahn was about to embark on represented far more than resolving the question of whether human beings had the power to affect inanimate objects, whether dice, spoons or microprocesses. This was study into the very nature of reality and the nature of living consciousness. This was science at its most wondrous and elemental.

Schmidt had taken great care to find special people with exceptional abilities who might be able to get especially good results. Schmidt's was a protocol of the extraordinary – abnormal feats performed by abnormal people with a peculiar gift. Jahn believed that this approach further marginalized the topic. The more interesting question, in his mind, was whether this was a capacity present in every human being.

He also wondered what impact this might have on our everyday lives. From his position as dean of an engineering school in the 1970s, Jahn realized that the world stood poised on the brink of a major computer revolution. Microprocessor technology was becoming increasingly sensitive and vulnerable. If it were true that living consciousness could influence such sensitive equipment, this in itself would have a major impact on how the equipment operated. The tiniest disturbances in a quantum process could create significant deviations from established behavior, the slightest movement send it soaring in a completely different direction.

Jahn knew that he was in a position to make a unique contribution. If this research were grounded in traditional science backed by a prestigious university, the entire topic might be aired in a more scholarly way.

He made plans for setting up a small program, and gave it a neutral name: Princeton Engineering Anomalies Research, which would thereafter always be known as PEAR. Jahn also resolved to take a low-key and lone-wolf approach by deliberately distancing himself from the various parapsychological associations and studiously avoiding any publicity.

Before long, private funding began rolling in, launching a precedent that Jahn would follow thereafter of never taking a dime of the University's money for his PEAR work. Largely because of Jahn's reputation, Princeton tolerated PEAR like a patient parent with a precocious but unruly child. He was offered a tiny cluster of rooms in the basement of the engineering school, which was to exist as its own little universe within one of the more conservative disciplines on this American Ivy League campus.

As Jahn began considering what he might need to get a program of this size off the ground, he made contact with many of the other new explorers in frontier physics and consciousness studies. In the process, he met and hired Brenda Dunne, a developmental psychologist at the University of Chicago, who had conducted and validated a number of experiments in clairvoyance.

In Dunne, Jahn had deliberately chosen a counterpoint to himself, which was obvious at first sight by their gaping physical differences. Jahn was spare and gaunt, often neatly turned out in a tidy checked shirt and casual trousers, the informal uniform of conservative academia, and in both his manner and his erudite speech gave off a sense of containment – never a superfluous word or unnecessary gesture. Dunne had the more effusive personal style. She was often draped in flowing clothes, her immense mane of salt-and-pepper hair hung loose or pony-tailed like a Native American. Although also a seasoned scientist, Dunne tended to lead from the instinctive. Her job was to provide the more metaphysical and subjective understanding of the material to bolster Jahn's largely analytical approach. He would design the machines; she would design the look and feel of the experiments. He would represent PEAR's face to the world; she would represent a less formidable face to its participants.

The first task, in Jahn's mind, was to improve upon the RNG technology. Jahn decided that his Random Event Generators, or REGs (hard 'G'), as they came to be called, should be driven by an electronic noise source, rather than atomic decay. The random output of these machines was controlled by something akin to the white noise you hear when the dial of your radio is between stations – a tiny roaring surf of free electrons. This provided a mechanism to send out a randomly alternating string of positive and negative pulses. The results were displayed on a computer screen and then transmitted on-line to a data management system. A number of failsafe features, such as voltage and thermal monitors, guarded against tampering or breakdown, and they were checked religiously to ensure that when not involved in experiments of volition, they were producing each of their two possibilities, 1 or 0, more or less 50 per cent of the time.

All the hardware failsafe devices guaranteed that any deviation from the normal 50–50 chance heads and tails would not be due to any electronic glitches, but purely the result of some information or influence acting upon it. Even the most minute effects could be quickly quantified by the computer. Jahn also souped up the hardware, getting it to work far faster. By the time he was finished, it occurred to him that in a single afternoon he could collect more data than Rhine had amassed in his entire lifetime.

Dunne and Jahn also refined the scientific protocol. They decided that all their REG studies should follow the same design: each participant sitting in front of the machine would undergo three tests of equal length. In the first, they would will the machine to produce more 1s then 0s (or 'HI's, as PEAR researchers put it). In the second, they would mentally direct the machine to produce more 0s than 1s (more 'LO's). In the third, they would attempt not to influence the machine in any way. This three-stage process was to guard against any bias in the equipment. The machine would then record the operator's decisions virtually simultaneously.

When a participant pressed a button, he would set off a trial of 200 binary 'hits' of 1 or 0, lasting about one-fifth of a second, during which time he would hold his mental intention (to produce more than the 100 '1's, say, expected by chance). Usually the PEAR team would ask each operator to carry out a run of 50 trials at one go, a process that might only take half an hour but which would produce 10,000 hits of 1 or 0. Dunne and Jahn typically examined scores for each operator of blocks of 50 or 100 runs (2,500 to 5,000 trials, or 500,000 to one million binary 'hits') – the minimum chunk of data, they determined, for reliably pinpointing trends.[17]

From the outset it was clear that they needed a sophisticated method of analyzing their results. Schmidt had simply counted up the number of hits and compared them to chance. Jahn and Dunne decided to use a tried-and-tested method in statistics called cumulative deviation, which entailed continually adding up your deviation from the chance score – 100 – for each trial and averaging it, and then plotting it on a graph.

The graph would show the mean, or average, and certain standard deviations – margins where results deviate from the mean but are still not considered significant. In trials of 200 binary hits occurring randomly, your machine should throw an average of 100 heads and 100 tails over time – so your bell curve will have 100 as its mean, represented by a vertical line initiated from top of its highest point. If you were to plot each result every time your machine conducted a trial, you would have individual points on your bell curve – 101, 103, 95, 104 – representing each score. Because any single effect is so tiny, it is difficult, doing it that way, to see any overall trend. But if you continue to add up and average your results and are having effects, no matter how slight, your scores should lead to a steadily increasing departure from expectation. Cumulative averaging shows off any deviation in bold relief.[18]

It was also clear to Jahn and Dunne that they needed a vast amount of data. Statistical glitches can occur even with a pool of data as large as 25,000 trials. If you are looking at a binary chance event like coin tossing, in statistical terms you should be throwing heads or tails roughly half the time. Say you decided to toss a coin 200 times and came up with 102 heads. Given the small numbers involved, your slight favouring of heads would still be considered statistically well within the laws of chance.

But if you tossed that same coin 2 million times, and you came up with 1,020,000 heads, this would suddenly represent a huge deviation from chance. With tiny effects like the REG tests, it is not individual or small clusters of studies but the combining of vast amounts of data which 'compounds' to statistical significance, by its increasing departure from expectation.[19]

After their first 5000 studies Jahn and Dunne decided to pull off the data and compute what was happening thus far. It was a Sunday evening and they were at Bob Jahn's house. They took their average results for each operator and began plotting them on a graph, using little red dots for any time their operators had attempted to influence the machine to have a HI (heads) and little green dots for the LO intentions (tails).

When they finished, they examined what they had. If there had been no deviation from chance, the two bell curves would be sitting right on top of the bell curve of chance, with 100 as the mean.

Their results were nothing like that. The two types of intention had each gone in a different direction. The red bell curve, representing the 'HI' intentions, had shifted to the right of the chance average, and the green bell curve had shifted to the left. This was as rigorous a scientific study as they come, and yet somehow their participants – all ordinary people, no psychic superstars among them – had been able to affect the random movement of machines simply by an act of will.

Jahn looked up from the data, sat back in his chair and met Brenda's eye. 'That's very nice,' he said.

Dunne stared at him in disbelief. With scientific rigor and technological precision they had just generated proof of ideas that were formerly the province of mystical experience or the most outlandish science fiction. They'd proved something revolutionary about human consciousness. Maybe one day this work would herald a refinement of quantum physics. Indeed, what they had in their hands was *beyond* current science – was perhaps the beginnings of a new science.

'What do you mean, "*that's very nice*"?' she replied. 'This is absolutely . . . *incredible*!'

Even Bob Jahn, in his cautious and deliberate manner, his dislike of being immoderate or waving a fist in the air, had to admit, staring at the graphs sprawled across his dining-room table, that there were no words in his current scientific vocabulary to explain them.

It was Brenda who first suggested that they make the machines more engaging and the environment more cosy in order to encourage the 'resonance' which appeared to be occurring between participants and their machines. Jahn began creating a host of ingenious random mechanical, optical and electronic devices – a swinging pendulum; a spouting water fountain; computer screens which switched attractive images at random; a moveable REG which skittled randomly back and forth across a table; and the jewel in the PEAR lab's crown, a random mechanical cascade. At rest it appeared like a giant pinball machine attached to the wall, a 6-by-10-foot framed set of 330 pegs. When activated, nine thousand polystyrene balls tumbled over the pegs in the span of only 12 minutes and stacked in one of nineteen collection bins, eventually producing a configuration resembling a bell-shaped curve. Brenda put a toy frog on the moveable REGs and spent time selecting attractive computer images, so that participants would be 'rewarded' if they chose a certain image by seeing more of it. They put up wood paneling. They began a collection of teddy bears. They offered participants snacks and breaks.

Year in and year out, Jahn and Dunne carried on the tedious process of collecting a mountain of data – which would eventually turn into the largest database ever assembled of studies into remote intention. At various points, they would stop to analyze all they had amassed thus far. In one 12-year period of nearly 2.5 million trials, it turned out that 52 per cent of all the trials were in the intended direction and nearly two-thirds of the ninety-one operators had overall success in influencing the machines the way they'd intended. This was true, no matter which type of machine was used.[20] Nothing else – whether it was the way a participant looked at a machine, the strength of their concentration, the lighting, the background noise or even the presence of other people – seemed to make any difference to the results. So long as the participant willed the machine to register heads or tails, he or she had some influence on it a significant percentage of the time.

The results with different individuals would vary (some would produce more heads than tails, even when they had concentrated on the exact opposite). Nevertheless, many operators had their own 'signature' outcome – Peter would tend to produce more heads than tails, and Paul vice versa.[21] Results also tended to be unique to the individual operator, no matter what the machine. This indicated that the process was universal, not one occurring with only certain interactions or individuals.

In 1987, Roger Nelson of the PEAR team and Dean Radin, both doctors of psychology, combined all the REG experiments – more than 800 – that had been conducted up to that time.[22] A pooling together of the results of the individual studies of sixty-eight investigators, including Schmidt and the PEAR team, showed that participants could affect the machine so that it gives the desired result about 51 per cent of the time, against an expected result of 50 per cent. These results were similar to those of two earlier reviews and an overview of many of the experiments performed on dice.[23] Schmidt's results remained the most dramatic with those studies that had leapt to 54 per cent.[24]

Although 51 or 54 per cent doesn't sound like much of an effect, statistically speaking it's a giant step. If you combine all the studies into what is called a 'meta-analysis', as Radin and Nelson did, the odds of this overall score occurring are a trillion to one.[25] In their meta-analysis, Radin and Nelson even took account of the most frequent criticisms of the REG studies concerning procedures, data or equipment by setting up sixteen criteria by which to judge each experimenter's overall data and then assigning each experiment a quality score.[26] A more recent meta-analysis of the REG data from 1959 to 2000 showed a similar result.[27] The US National Research Council also concluded that the REG trials could not be explained by chance.[28]

An effect size is a figure which reflects the actual size of change or outcome in a study. It is arrived at by factoring in such variables as the number of participants and the length of the test. In some drug studies, it is arrived at by dividing the number of people who have had a positive effect from the drug by the total number of participants in the trial. The overall effect size of the PEAR database was 0.2 per hour.[29] Usually an effect size between 0.0 to 0.3 is considered small, a 0.3 to 0.6 effect size is medium and anything above that is considered large. The PEAR effect sizes are considered small and the overall REG studies, small to medium. However, these effect sizes are far larger than those of many drugs deemed to be highly successful in medicine.

Numerous studies have shown that propranolol and aspirin are highly successful in reducing heart attacks. Aspirin in particular has been hailed as a great white hope of heart disease prevention. Nevertheless, large studies have shown that the effect size of propranolol is 0.04 and aspirin is 0.03, respectively – or about ten times smaller than the effect sizes of the PEAR data. One method of determining the magnitude of effect sizes is to convert the figure to the number of persons surviving in a sample of 100 people. An effect size of 0.03 in a medical life-or-death situation would mean that three additional people out of one hundred survived, and an effect size of 0.3 would mean that an additional thirty of one hundred survived.[30]

To give some hypothetical idea of the magnitude of the difference, say that with a certain type of heart operation, thirty patients out of a hundred usually survive. Now, say that patients undergoing this operation are given a new drug with an effect size of 0.3 – close to the size of the hourly PEAR effect. Offering the drug on top of the operation would virtually double the survival rate. *An additional effect size of 0.3 would turn a medical treatment that had been life-saving less than half the time into one that worked in the majority of cases.*[31]

Other investigators using REG machines discovered that it was not simply humans who had this influence over the physical world. Using a variation of Jahn's REG machines, a French scientist named René Peoc'h also carried out an ingenious experiment with baby chicks. As soon as they were born, a moveable REG was 'imprinted' on them as their 'mother'. The robot was then placed outside the chicks' cage and allowed to move about freely, as Peoc'h tracked its path. After a time, the evidence was clear – the robot was moving toward the chicks more than it would do if it were wandering randomly. The desire of the chicks to be near their mother was an 'inferred intention' that appeared to be having an effect in drawing the machine nearer.[32] Peoc'h carried out a similar study with baby rabbits. He placed a bright light on the moveable REG that the baby rabbits found abhorrent. When the data from the experiment were analyzed, it appeared that the rabbits were successfully willing the machine to stay away from them.

Jahn and Dunne began to formulate a theory. If reality resulted from some elaborate interaction of consciousness with its environment, then consciousness, like subatomic particles of matter, might also be based on a system of probabilities. One of the central tenets of quantum physics, first proposed by Louis de Broglie, is that subatomic entities can behave either as particles (precise things with a set location in space) or waves (diffuse and unbounded regions of influence which can flow through and interfere with other waves). They began to chew over the idea that consciousness had a similar duality. Each individual consciousness had its own 'particulate' separateness, but was also capable of 'wave-like' behavior, in which it could flow through any barriers or distance, to exchange information and interact with the physical world. At certain times, subatomic consciousness would get in resonance with – beat at the same frequency as – certain subatomic matter. In the model they began to assemble, consciousness 'atoms' combined with ordinary atoms – those, say, of the REG machine – and created a 'consciousness molecule' in which the whole was different from its component parts. The original atoms would each surrender their individual entities to a single larger, more complex entity. On the most basic level, their theory was saying, you and your REG machine develop coherence.[33]

Certainly some of their results seemed to favor this interpretation. Jahn and Dunne had wondered if the tiny effect they were observing with individuals would get any larger if two or more people tried to influence the machine in tandem. The PEAR lab ran a series of studies using pairs of people, in which each pair was to act in concert when attempting to influence the machines.

Of 256,500 trials, produced by fifteen pairs in forty-two experimental series, many pairs also produced a 'signature' result, which didn't necessarily resemble the effect of either individual alone.[34] Being of the same sex tended to have a very slight negative effect. These types of couples had a worse outcome than they achieved individually; with eight pairs of operators the results were the very opposite of what was intended. Couples of the opposite sex, all of whom knew each other, had a powerful complementary effect, producing more than three and a half times the effect of individuals. However, 'bonded' pairs, those couples in a relationship, had the most profound effect, which was nearly six times as strong as that of single operators.[35]

If these effects depended upon some sort of resonance between the two participating consciousnesses, it would make sense that stronger effects would occur among those people sharing identities, such as siblings, twins or couples in a relationship.[36] Being close may create coherence. As two waves in phase amplified a signal, it may be that a bonded couple has an especially powerful resonance, which would enhance their joint effect on the machine.

A few years later, Dunne analyzed the database to see if results differed according to gender. When she divided results between men and women, she found that men on the whole were better at getting the machine to do what they wanted it do, although their overall effect was weaker than it was with women. Women, on the whole, had a stronger effect on the machine, but not necessarily in the direction they'd intended.[37] After examining 270 databases produced by 135 operators in nine experiments between 1979 and 1993, Dunne found that men had equal success in making the machine do what they wanted, whether heads or tails (or HIs and LOs). Women, on the other hand, were successful in influencing the machine to record heads (HIs), but not tails (LOs). In fact, most of their attempts to get the machine to do tails failed. Although the machine would vary from chance, it would be in the very opposite direction of what they'd intended.[38]

At times, women produced better results when they weren't concentrating strictly on the machine, but were doing other things as well, whereas strict concentration seemed important for men's success.[39] This may provide some subatomic evidence that women are better at multitasking than men, while men are better at concentrated focus. It may well be that in microscopic ways men have a more direct impact on their world, while women's effects are more profound.

Then something happened which forced Jahn and Dunne to reconsider their hypothesis about the nature of the effects they were observing. In 1992, PEAR had banded together with the University of Giessen and the Freiberg Institute to create the Mind–Machine Consortium. The consortium's first task was to replicate the original PEAR data, which everyone assumed would proceed as a matter of course. Once the results of all three laboratories were examined, however, they looked, at first glance, a failure – little better than the 50–50 odds which occur by chance alone.[40]

When writing up the results, Jahn and Dunne noticed some odd distortions in the data. Something interesting had occurred in the secondary variables. In statistical graphs, you can show not only what your average ought to be but also how far the deviations from it ought to spread from your mean. With the Mind–Machine data, the mean was right where it would be with a chance result, but not much else was. The size of the variation was too big, and the shape of the bell curve was disproportionate. Overall, the distribution was far more skewed than it would be if it were just a chance result. Something strange was going on.

When Jahn and Dunne looked a little closer at the data, the most obvious problem had to do with feedback. Up until that time they'd operated on the assumption that providing immediate feedback – telling the operators how they were doing in influencing the machine – and making an attractive display or a machine that people could really engage with would crucially help to produce good results. This would hook the operator into the process and help them to get in 'resonance' with the device. For the mental world to interact with the physical world, they'd thought, the interface – an attractive display – was crucial in breaching that divide.

However, in the Consortium data, they realized that the operators were doing just as well – or sometimes better – when they had no feedback.

One of their other studies, called ArtREG, had also failed to get significant overall results.[41] They decided to examine that study a bit more closely in light of the Mind–Machine Consortium results. They'd used engaging images on a computer, which randomly switched back and forth – in one case a Navajo sand painting switched with Anubis, the ancient Egyptian judge of the dead. The idea was for their operators to will the machine to show more of one than the other. The PEAR team had assumed once again that an attractive image would act as a carrot – you'd be 'rewarded' for your intention by seeing more of the image you preferred.

Once they'd examined the data of the study in terms of yield by picture, those images which had produced the most successful outcomes all fell into a similar category: the archetypal, the ritualistic or the religiously iconographic. This was the domain of dreams, the unexpressed or unarticulated – images that, by their very design, were intended to engage the unconscious.

If that were true, the intention was coming from deep in the unconscious mind, and this may have been the cause of the effects. Jahn and Dunne realized what was wrong with their assumptions. Using devices to make the participant function on a conscious level might be acting as a barrier. Instead of increasing conscious awareness among their operators, they should be diminishing it.[42]

This realization caused them to refine their ideas about how the effects they'd observed in their labs might occur. Jahn liked to call it his 'work in progress'. It appeared that the unconscious mind somehow had the capability of communicating with the subtangible physical world – the quantum world of all possibility. This marriage of unformed mind and matter would then assemble itself into something tangible in the manifest world.[43]

This model makes perfect sense if it also embraces theories of the Zero Point Field and quantum biology proposed by Pribram, Popp and the others. Both the unconscious mind – a world before thought and conscious intention – and the 'unconscious' of matter – the Zero Point Field – exist in a probabilistic state of all possibility. The subconscious mind is a preconceptual substrate from which concepts emerge, and the Zero Point Field is a probabilistic substrate of the physical world. It is mind and matter at their most fundamental. In this subtangible dimension, possibly of a common origin, it would make sense that there would be a greater likelihood of quantum interaction.

At times, Jahn kicked around the most radical idea of all. When you get down far enough into the quantum world, there may be no distinction between the mental and the physical. There may be only the concept. It might just be consciousness attempting to make sense of a blizzard of information. There might not be two intangible worlds. There might be only one – The Field and the ability of matter to organize itself coherently.[44]

As Pribram and Hameroff theorized, consciousness results from superradiance, a rippling cascade of subatomic coherence – when individual quantum particles such as photons lose their individuality and begin acting as a single unit, like an army calling every soldier into line. Since every motion of every charged particle of every biological process is mirrored in the Zero Point Field, our coherence extends out in the world. According to the laws of classical physics, particularly the law of entropy, the movement of the inanimate world is always toward chaos and disorder. However, the coherence of consciousness represents the greatest form of order known to nature, and the PEAR studies suggest that this order may help to shape and create order in the world. When we wish for something or intend something, an act which requires a great deal of unity of thought, our own coherence may be, in a sense, infectious.

On the most profound level, the PEAR studies also suggest that reality is created by each of us *only by our attention*. At the lowest level of mind and matter, each of us creates the world.

The effects that Jahn had been able to record were almost imperceptible. It was too early to know why. Either the machinery was still too crude to pick up the effect or he was only picking up a single signal, when the real effect occurs from an ocean of signals – an interaction of all living things in the Zero Point Field. The difference between his own results and the higher ones recorded by Schmidt suggested that this ability was spread across the population, but that it was like artistic ability. Certain individuals were more skillful at harnessing it.

Jahn had seen that this process had minute effects on probabilistic processes, and that this might explain all the well-known stories about people having positive or negative effects on machines – why, on some bad days, computers, telephones and photocopiers malfunction. It might even explain the problems Benveniste had been having with his robot.

It seemed that we had an ability to extend our own coherence out into our environment. By a simple act of wishing, we could create order. This represented an almost unimaginable amount of power. On the crudest level, Jahn had proved that, at least on the subatomic level, there was such as thing as mind over matter. But he'd demonstrated something even more fundamental about the powerful nature of human intention. The REG data offered a tiny window into the very essence of human creativity – its capacity to create, to organize, even to heal.[45] Jahn had his evidence that human consciousness had the power to order random electronic devices. The question now before him was what else might be possible.

CHAPTER SEVEN

Sharing Dreams

DEEP IN THE RAINFORESTS of the Amazon, the Achuar and the Huaorani Indians are assembled for their daily ritual. Every morning, each member of the tribe awakens before dawn, and once gathered together in that twilight hour, as the world explodes into light, they share their dreams. This is not simply an interesting pastime, an opportunity for storytelling: to the Achuar and the Huaorani, the dream is owned not by the dreamer alone, but collectively by the group, and the individual dreamer is simply the vessel the dream decided to borrow to have a conversation with the whole tribe. The tribes view the dream as a map for their waking hours. It is a forecaster of what is to come for all of them. In dreams they connect with their ancestors and the rest of the universe. The dream is what is real. It is their waking life that is the falsehood.[1]

Further north, a group of scientists also discovered that dreams aren't owned by the dreamer, asleep in a soundproof chamber behind an electromagnetic shield, electrodes taped to his skull. They are owned by Sol Fieldstein, a City College doctoral student in another room several hundred yards away, who is examining a painting entitled *Zapatistas* by Carlos Orozco Romero – a panorama of Mexican revolutionaries, followers of Emiliano Zapata, all marching with their shawled women under the dark clouds of an imminent storm. Sol's instructions are to will this image to the dreamer. A few moments later, the dreamer, Dr William Erwin, a psychoanalyst, is awakened. The dream he was having, he told them, was a crazy thing, almost like a colossal Cecil B. DeMille production. What he kept seeing was this image, under a foreboding sky, of some sort of ancient Mexican civilization.[2]

The dreamer is the vessel for a borrowed thought, a collective notion, present in the microscopic vibrations in between the dreamers. The dream state is more authentic for it shows the connection in bold relief. Their waking state of isolation, each in their separate room, is, as the Amazons view it, the impostor.

One of the questions that arose from the PEAR studies was the nature of ownership of thought. If you could influence machines, it rather begged the question of exactly where your thoughts lie. Where exactly was the human mind? The usual assumption in Western culture is that it is located in our brains. But if this is true, how could thoughts or intentions affect other people? Is it that the thought is 'out there', somewhere else? Or is there such a thing as an extended mind, a collective thought? Does what we think or dream influence anyone else?

These were the kinds of questions that preoccupied William Braud. He'd read of studies like the one with the Mexican painting, which was one of the more dramatic of studies on telepathy conducted by Charles Honorton, a noted consciousness researcher at the Maimonides Medical Center in Brooklyn, New York. For a behaviorist like Braud, the Honorton study represented a radical new education.

Braud was soft-spoken and thoughtful, with a gentle, deliberate manner, most of his face encompassed by a generous beard. He'd begun his career as a psychologist of the old school, with a particular interest in the psychology and biochemistry of memory and learning. Nevertheless, there was an errant streak in him, a fascination with what William James, the founder of psychology in America, had termed 'white crows'. Braud liked anomalies, the things in life that didn't fit, the assumptions that could be turned askew.

Just a few years after he'd got his PhD, the 1960s had loosened up the tight hold of Pavlov and Skinner on his imagination. At the time, Braud had been teaching classes in memory, motivation and learning at the University of Houston. Recently, he'd become interested in work showing a remarkable property of the human brain. The early pioneers in biofeedback and relaxation demonstrated that people could influence their own muscular reaction or heart rate, just by directing their attention to parts of it in sequence. Biofeedback even had measurable effects on brain wave activity, blood pressure and electrical activity on the skin.[3]

Braud had been toying with his own studies on extrasensory perception. One of his students who practiced hypnosis agreed to participate in a study in which Braud attempted to transmit his thoughts. Some amazing transferences had gone on. His student, who'd been hypnotized and was sitting in a room down the hall from him, unaware of Braud's doings, seemed to have some empathetic connection with him. Braud had pricked his hand and placed it over a candle flame and his student experienced pain or heat. He'd looked at a picture of a boat and the student remarked about a boat. He opened the door of his lab into the brilliant Texas sunshine and the student mentioned the sun. Braud had been able to carry out his end of the experiment anywhere – the other side of the building or many miles away from his student in the sealed room – and get the same results.[4]

In 1971, when he was 29, Braud crossed paths with Edgar Mitchell, who had just returned from his *Apollo 14* flight. Mitchell had decided to write a book about the nature of consciousness and at the time he was scouting around for any good research of this kind. Braud and one other academic were the only people in Houston involved in any credible study of the nature of consciousness. It was only natural that he and Mitchell would find each other. They began meeting regularly and comparing notes on research that existed in this area.

There was plenty of research on telepathy. There'd been the highly successful card experiments of Joseph Rhine, used by Mitchell in outer space. Even more convincing were the studies of the Maimonides Medical Center in Brooklyn in the late 1960s, conducted in its special dream research laboratory. Montague Ullman and Stanley Krippner had conducted numerous experiments like the one with the Mexican painting to see if thoughts could be sent and incorporated into dreams. The Maimonides work had been so successful[5] that when analyzed by a University of California statistician who was expert in psychic research, the total series had showed an astonishing accuracy rate of 84 per cent. The odds of this happening by chance were a quarter of a million to one.[6]

There'd even been some evidence that people can empathetically feel another's pain. A psychologist named Charles Tart in Berkeley had designed a particularly brutal study, administering electric shocks to himself to see if he could 'send' his pain and have it register with a receiver, who was hooked up to machines which would measure heart rate, blood volume and other physiological changes.[7] What Tart found was that his receivers were aware of his pain, but not on a conscious level. Any empathy they might have had was registering physiologically through decreased blood volume or faster beating of the heart – but not consciously. When questioned, the participants hadn't any idea when Tart was getting the shocks.[8]

Tart also had shown that when two participants hypnotize each other, they experience intense common hallucinations. They also claimed to have shared an extrasensory communication, where they knew each other's thoughts and feelings.[9]

It got so that Braud's white crows were beginning to take over, crowding out his academic work. Braud's own belief system had moved in small deliberate steps from his original ideas, which had embraced the simple cause-and-effect equations of brain chemistry, to more complex ideas about consciousness. His own tentative experiments had been so breathtakingly dramatic that they had convinced him that something far more complex than chemicals was at work in the brain – if any of this was happening in the brain at all.

As he'd become interested in altered consciousness and the effect of relaxation on physiology, so Braud had been lured away from his behaviorist theories. Mitchell had been receiving some funding from the Mind Science Foundation, an organization devoted to consciousness research. As it happened, the Foundation was planning to move to San Antonio and needed another senior scientist. The job, with all the freedom it offered for experimentation into the nature of consciousness, was exactly what Braud was looking for.

The world of consciousness research was a small one. One of the other members of the Foundation was Helmut Schmidt, and Braud soon met Schmidt and his REG machines. It was there that he began to wonder how far the influence of the human mind worked. After all, human beings, like REGs, qualify as systems with considerable plasticity and lability – potential for change. These dynamic systems were always in flux and might also be susceptible to psychokinetic influence on some level – quantum or otherwise.

It was only one small step further for Braud to consider that if people could affect their own bodies through attention, then they just might be able to create the same effect in someone else. And if we could create order in inanimate objects such as REG machines, perhaps we could also establish order in other living things. What these thoughts were leading up to was a model of consciousness that was not even limited by the body, but was an ethereal presence that trespassed into other bodies and living things and affected them as if they were its own.

Braud decided to develop a series of experiments to explore just how much influence individual intention might have on other living things. These were difficult studies to design. The problem with most living systems is their sheer dynamism. There are so many variables that it is hard to measure change. Braud decided to begin with simple animals and slowly advance in evolutionary complexity. He needed a simple system with some capability of changing in easily measurable ways. Research of his chanced upon a perfect candidate. He discovered that the small knife fish (*Gymnotus carapo*) emits a weak electrical signal, which is probably used for navigational purposes. The electrical signal would allow him to quantify its direction precisely. Electrodes fastened to the side of a small tank would pick up the electrical activity of the fish's emissions and give an influencer immediate feedback on an oscilloscope screen. The question was whether people could change the fish's swimming orientation.

Mongolian gerbils were another good candidate because they like to run in activity wheels. This also gave Braud something to measure. He could quantify the velocity of a gerbil on its run and then see if human intention could make it go faster.

Braud wanted to test the effects of intention on human cells, ideally those of the immune system, for if an outside agent could influence the immune system, the prospects for healing were immense. But this represented a challenge far too great for his laboratory. The immune system was an entity with so much complexity that in any study of human intention, it would be almost impossible to quantify what had changed and who was responsible for the changing.

A far better candidate was the red blood cell. When red blood cells are placed in a solution with the same saline (salt) levels as blood plasma, their membranes remain intact and will survive for a long time. Add too much or too little salt to the solution and the membranes of the blood cells weaken and finally burst, causing the hemoglobin of the cell to spill out into the solution, a process called 'hemolysis'. Controlling the rate is often a matter of varying the amount of salt in the solution. Since the solution becomes more transparent as hemolysis carries on, you can also quantify the rate of this process by measuring the amount of light transmitted through the solution with a gadget called a spectrophotometer. Here was another system which was easy to measure. Braud decided to enlist some volunteers, place them in a distant room and determine whether, by simple wishing, they could 'protect' these cells from bursting by slowing their rate of hemolysis once a fatal amount of salt had been added into the test tube.

All these studies met with success.[10] Braud's volunteers had been able to change the direction of fish, speed up gerbils and protect human red blood cells to a significant extent. Braud was ready to move on to human beings, but he needed some method of isolating physical effects. A perfect device for this, as any police detective knows, is one that measures electrodermal activity (EDA). With lie detector tests, the machine picks up any increase in the electrical conductivity of the skin, which is caused by increased activity of the sweat glands, which in turn are governed by the sympathetic nervous system. As doctors can measure electrical activity of the heart and brain with ECG (electrocardiogram) and (EEG) electroencephalogram) machines, respectively, so too can the lie detector record increased electrodermal activity. Higher EDA readings show that the sympathetic nervous system, which governs emotional states, is in overdrive. This would indicate stress, emotion or mood swings – any sort of heightened arousal – which is more likely if someone is lying. These are often referred to as 'fight or flight' responses, which rise and become more pronounced when we face something dangerous or upsetting: our hearts race, our pupils dilate, our skin tends to sweat more and blood drains from our extremities to go to the sites in the body where it is most needed. Taking these readings can give you a measure of unconscious response, when the sympathetic nervous system is stressed before the person being tested is even consciously aware of it. By the same token, low levels of EDA would be indicative of little stress and a state of calm – the natural state of truth telling.

Braud launched his human experimentation with what would become one of his signature studies: the effect of being stared at. Researchers into the nature of consciousness are particularly fond of the phenomenon because it is a relatively easy extrasensory experiment with which to judge success. With transmitted thoughts, there are many variables to consider when determining whether the receiver's response matches the sender's thoughts. With staring, the receiver either feels it or doesn't. It is the closest you can get to reducing subjective feelings to the simple binary multiple choice of a REG machine.

In Braud's hands, staring and being stared at became state of the art, a stalker's paradise. Participants would be placed in a room and be attached to silver chloride palmar electrodes, a skin resistance amplifier and a computer. The only other equipment in the room was a Hitachi color Camcorder VM-2250, which was to be the implement of spying. This small video camera would be attached to a 19-inch Sony Trinitron in another room, two hallways and four doors away. This would allow the starer to view the subject peacefully without the possibility of any form of sensory cueing.

Pure chance, as arrived at by artful mathematical calculation – a computer's random algorithm – governed the starer's script. Whenever the script dictated, the starer would stare intently at the subject on the monitor and attempt to gain his or her attention. Meanwhile, in the other room, the staree, relaxed in a reclining chair, had been told to think about anything other than wondering when he or she was being stared at.

Braud carried out this experiment sixteen times. In most cases, those being stared at showed significantly greater electrodermal activity during the staring sessions than would be expected by chance (59 per cent against the expected 50 per cent) – even though they were not consciously aware of it. With his second group of participants, Braud decided to try something different. In this case, he had them meet each other beforehand. He asked them to carry out a series of exercises that involved staring into each other's eyes and looking intently at each other when they talked. The idea was to reduce any discomfort over being stared at and also to get them to know each other. When this group underwent the trial, they got opposite results from the earlier tests. They were at their calmest precisely when they were being stared at. Like the Stockholm Syndrome, a psychological condition where prisoners begin to love their jailers, the starers had begun to love being stared at. In a manner of speaking, they'd become addicted to it. They were more relaxed when being stared at, even at a distance, and they missed it when no one was looking at them.[11]

From these latest studies, Braud grew even more convinced that people had some means of communicating and responding to remote attention, even when they weren't aware of it.[12] Like those people given Charles Tart's electric shocks, the person being stared at was not conscious of any of this. Awareness occurred only deep in a subliminal level.

Much of this research inspired an important consideration – the degree to which necessity dictated the size of the effect. It was obvious now to Braud that random systems or those with a high potential for influence could be affected by human intention. But was the effect any larger if the system *needed* changing? If it was possible to calm someone down, would the effect be more exaggerated in someone who needed calming down – someone, say, with loads of nervous energy? In other words, did *need* allow someone greater access to effects from The Field? Were the more organized of us – biologically speaking – better at accessing this information and drawing it to the attention of others?

In 1983, Braud tested out this theory with a series of studies in collaboration with an anthropologist called Marilyn Schlitz, another consciousness researcher who'd worked with Helmut Schmidt. Braud and Schlitz selected a group of highly nervous people, as evidenced by high sympathetic nervous system activity, and another calmer group. Using a similar protocol to the staring studies, Braud and Schlitz by turns tried to calm down members of both groups. Success or failure would be measured again by a polygraph tracing of a person's electrodermal activity.

The volunteers were also asked to participate in another experiment, in which they'd attempt to calm themselves down with standard relaxation methods.

When they finished the study, Schlitz and Braud noticed a huge disparity between results of the two groups.[13] As they suspected, the effect was far larger in the group needing the calming down. In fact, it was the greatest effect achieved in any of Braud's studies. The calm group, on the other hand, had registered almost no change; their effect only differed slightly from chance.

Strangest of all, the size of the effect on the agitated group by those trying to calm them down was only slightly less than the effect that people had on themselves when using relaxation techniques. In statistical terms, it meant that other people could have almost the same mind–body effect on you that you could have on yourself. Letting someone else express a good intention for you was almost as good as using biofeedback on yourself.

Braud tried a similar study showing that you could also help someone else focus his or her attention by remote influence. Once again, the effects were largest among those whose attention seemed to wander the most.[14]

A meta-analysis is a scientific method of assessing whether an observed effect is real and significant by pooling the data from a large body of often disparate individual studies. In effect, it combines single studies, which may sometimes be discounted as too small to be definitive, into one giant experiment. Although there are problems comparing studies of different shapes and sizes, it may give you some idea about whether the effect you are studying is big or small. Schlitz and Braud had conducted a meta-analysis on all of the studies they could find investigating the effect of intention on other living things. Research conducted all over the world had shown that human intention could affect bacteria and yeast, plants, ants, chicks, mice and rats, cats and dogs, human cellular preparations and enzyme activity. Studies on humans had shown that one set of people could successfully affect the eye or gross motor movements, breathing and even the brain rhythms of another set. The effects were small, but they occurred consistently and had been achieved by ordinary people who had been recruited to try out this ability for the very first time.

Overall, according to Schlitz and Braud's meta-analysis, the studies had a success rate of 37 per cent against the expected result of 5 per cent by chance.[15] The EDA studies alone had a success rate of 47 per cent compared with the 5 per cent success rate expected by chance.[16]

These results gave Braud several important clues about the nature of remote influence. It was apparent that ordinary humans had the ability to influence other living things on many levels: muscle activity, motor activity, cellular changes, nervous system activity. One other strange possibility was suggested by all these studies: the influence increased depending on how much it mattered to the influencer, or how much he or she could relate to the object of influence. The smallest effects were found in the fish studies; these increased in experiments dealing with cuddly gerbils; they increased yet again with human cells; and they were at their greatest when people were attempting to influence another person. But the greatest effect of all occurred when the people to be influenced really needed it. Those who required something – calming down, focusing attention – seemed more receptive to influence than others. And strangest of all, your influence on others was only marginally less than your influence on yourself.

Braud had even seen cases of telepathy during the influence sessions. At the beginning of one session, one influencer happened to remark that the electrodermal tracings of the subject were so regimented that they reminded him of a German techno-pop musical band called Kraftwerk. When Braud returned to the recipient's room at the end of the session, the first thing she said was that early in the session, for some odd reason, she kept thinking of the pop group Kraftwerk. In Braud's work this kind of association was becoming the norm, rather than the exception.[17]

Every scientist engaged in consciousness research was thinking the same thought. Why was it that some people were more able to influence, and some conditions more conducive to influence, than others? It was like a secret labyrinth that certain people could maneuver around more easily than others. Jahn and Dunne had found that archetypal or mythical images triggering the unconscious produced the strongest psychokinetic effects. The highly successful Maimonides research on telepathy had been conducted when the participants were asleep and dreaming. Even when only dabbling, Braud showed great success during hypnosis. In Tart's studies, and in his own remote staring studies, the communication had occurred subconsciously, without the recipient being aware of it.

Braud had looked hard for the common thread in all these experiments. He'd noticed several characteristics which tended to more readily guarantee success: some sort of relaxation technique (through meditation, biofeedback or another method); reduced sensory input or physical activity; dreams or other internal states and feelings; and a reliance on right-brain functioning.

Braud and others found what had been termed the 'sheep/goat' effect – these effects work better if you believe they will and less than average if you believe they won't. In each case, like a REG machine, you are affecting the result – even if (as a goat) your effect is negative.

Another important characteristic appeared to be an altered view of the world. People were more likely to succeed if, instead of believing in a distinction between themselves and the world, and seeing individual people and things as isolated and divisible, they viewed everything as a connected continuum of interrelations – and also if they understood that there were other ways to communicate than through the usual channels.[18]

It seemed that when the left brain was quieted and the right brain predominated, ordinary people could gain access to this information. Braud had read the *Vedas*, India's bible of the ancient Hindus, which described *siddhis*, or psychic events, that would occur during profound meditative states. In the highest state, the meditator experiences feelings of a type of omniscient knowing – a sense of seeing everywhere at once. The subject enters a state of unity with the single object being focused upon. He or she also experiences the ability to achieve gross psychokinetic effects such as levitation and moving objects at a distance.[19] In nearly every instance, the recipient had eliminated the sensory bombardment of the everyday and tapped into a deep well of alert receptivity.

Could it be that this communication is like any ordinary form of communication, but the noise of our everyday lives stops us hearing it? Braud realized that if he could create a state of sensory deprivation in a person, his mind might more readily notice the subtle effects not perceived by the ordinary chattering brain. Would perception improve if you deprived it of ordinary stimuli? Would this allow you access to The Field?

This was precisely the theory of Mahareshi Mahesh Yogi, the founder of Transcendental Meditation. Several studies carried out by the Moscow Brain Research Institute's Laboratory of Neurocybernetics examining the effect of TM on the brain show an increase in areas of the cortex taking part in the perception of information and also an increase in the functioning relationship of the left and right hemispheres of the brain. The studies would suggest that meditation opens the doors of perception a little wider.[20]

Braud had heard about the *ganzfeld*, which is German for 'whole field', a method of cutting out sensory input, and he began conducting ESP studies using a classic *ganzfeld* protocol. His volunteers would sit in a comfortable reclining chair in a soundproof room with soft lighting. Half spheres like halved ping-pong balls would be placed over their eyes and they would wear headphones, which played continuous, quiet static. Braud told the volunteers to speak for twenty minutes about any impressions that popped into their heads.

Thereafter, the study would follow the usual design of a telepathy experiment. Braud's hunch proved correct. The *ganzfeld* experiments were among the most successful of all.

When Braud's own studies were combined with twenty-seven others, twenty-three, or 82 per cent, were found to have success rates higher than chance. The median effect size was 0.32 – not dissimilar to PEAR's REG effect size.[21]

Important shifts in thinking often occur in interesting synchronicities. Charles Honorton of the Maimonides clinic in Brooklyn and Adrian Parker, a psychologist at the University of Edinburgh, had been wondering exactly the same thing as Braud and also began looking into the *ganzfeld* as a means of exploring the nature of human consciousness. The combined meta-analysis of all *ganzfeld* experiments produced a result with odds against chance of ten billion to one.[22]

Braud even experienced some premonitions when using the *ganzfeld* on himself. One evening, sitting on the floor of the living-room in his apartment in Houston, the half ping-pong balls and headphones in place, he suddenly experienced an intense and vivid vision of a motorcycle, with bright headlights and wet streets.

Soon after he'd finished his session, his wife returned home. At the very point he'd had his vision, she told him, she'd nearly collided with a motorcycle. There had been bright headlights shining at her and the streets were drenched with rain.[23]

Thoughts about the significance of his work percolated up in Braud's mind to a disquieting realization. If we could intend good things to happen to other people, we might also be able to make bad things happen.[24] There'd been many anecdotal stories of voodoo effects, and it made perfect sense, given the experimental results he'd been getting, that bad intentions could have an effect. Was it possible to protect yourself from them?

Some preliminary work of Braud's reassured him. One of his studies showed that it was possible for you to block or prevent any influences you didn't want.[25] This was possible through psychological 'shielding strategies'. You could visualize a safe or protective shield, or barrier or screen, which would prevent penetration of the influence.[26] In this experiment, participants were told to attempt to 'shield' themselves against the influence of two experimenters, who attempted to raise their EDA levels. The same was tried on another group, but they were told not to try to block any remote influence. Those doing the influencing weren't aware of who was blocking their attempts and who wasn't. At the end of the experiment, the shielded group showed far fewer physical effects than those who just allowed themselves to be affected.[27]

All the early ESP work had created a model of a mental radio, where one subject was sending thoughts to someone else. Braud now believed that the truth was far more complex. It appeared that the mental and physical structures of the sender's consciousness are able to exert an ordering influence on the less-organized recipient. Another possibility was that it was all there all the time, in some type of field, like the Zero Point Field, which could be tapped into and mobilized when necessary. This was the view of David Bohm, who'd postulated that all information was present in some invisible domain, or higher reality (the implicate order), but active information could be called up, like a fire brigade, at time of need, when it would be necessary and meaningful.[28] Braud suspected the answer might be a mixture of the latter two – a field of all information and an ability of human beings to provide information which would help to better order other people and things. In ordinary perception, the capacity of the dendritic networks in our brains to receive information from the Zero Point Field is strictly limited, as Pribram demonstrated. We are tuned in to only a limited range of frequencies. However, any state of altered consciousness – meditation, relaxation, the *ganzfeld*, dreams – relaxes this constraint. According to systems theorist Ervin Laszlo, it is as though we are a radio and our 'bandwidth' expands.[29] The receptive patches in our brains become more receptive to a larger number of wavelengths in the Zero Point Field.

Our ability to pick up signals also increases during the kind of deep interpersonal connection examined by Braud. When two people 'relax' their bandwidths and attempt to establish some kind of deep connection, their brain patterns become highly synchronized.

Studies in Mexico similar to Braud's, where a pair of volunteers in separate rooms were asked to feel each other's presence, showed that the brain waves of both participants, as measured by EEG readings, began to synchronize. At the same time, electrical activity within each hemisphere of the brain of each participant also synchronized, a phenomenon which usually only occurs in meditation. Nevertheless, it was the participant with the most cohesive brain-wave patterns who tended to influence the other. The most ordered brain pattern always prevailed.[30]

In this circumstance, a type of 'coherent domain' gets established, just as with molecules of water. The ordinary boundary of separateness is crossed. The brain of each member of the pair becomes less highly tuned in to their own separate information and more receptive to that of the other. In effect, they pick up someone else's information from the Zero Point Field as if it were their own.

As quantum mechanics govern living systems, quantum uncertainty and probability are features of all our bodily processes. We are walking REG machines. At any moment of our lives, any one of the microscopic processes that make up our mental and physical existence can be influenced to take one of many paths. In the circumstance of Braud's studies, in which two people have a 'synchronized' bandwidth, the observer with the greater degree of coherence, or order, influences the probabilistic processes of the less organized recipient. The more ordered of Braud's pairs affects some quantum state in the more disordered other and nudges it to toward a greater degree of order.

Laszlo believes that this notion of 'expanded' bandwidth would account for a number of puzzling and highly detailed reports of people who undergo regression therapy or claim to remember past lives, a phenomenon which mainly occurs among very young children.[31] EEG studies of the brains of children under five show that they permanently function in alpha mode – the state of altered consciousness in an adult – rather than the beta mode of ordinary mature consciousness. Children are open to far more information in The Field than the average adult. In effect, a child walks around in a state of a permanent hallucination. If a small child claims to remember a past life, the child might not be able to distinguish his own experiences from someone else's information, as stored in the Zero Point Field. Some common trait – a disability or special gift, say – might trigger an association, and the child would pick up this information as if it were his own past-life 'memory'. It is not reincarnation, but just accidentally tuning into somebody else's radio station by someone who has the capacity to receive a large number of stations at any one time.[32]

The model suggested by Braud's work is of a universe, to some degree, under our control. Our wishes and intentions create our reality. We might be able to use them to have a happier life, to block unfavorable influences, to keep ourselves enclosed in a protective fence of goodwill. Be careful what you wish for, thought Braud. Each of us has the ability to make it come true.

In his own casual and quiet way, Braud began testing out this idea, using intentions to achieve certain outcomes. It only seemed to work, he discovered, when he used gentle wishing, rather than intense willing or striving. It was like trying to will yourself to sleep: the harder you try, the more you interfere with the process. It seemed to Braud that humans operated on two levels – the hard, motivated striving of the world and the relaxed, passive, receptive world of The Field – and the two seemed incompatible. Over time, when Braud's desired outcomes seemed to occur more often than expected by chance, he developed a reputation as a 'good wisher.'[33]

Braud's work offered further proof of what many other scientists were beginning to realize. Our natural state of being is a relationship – a tango – a constant state of one influencing the other. Just as the subatomic particles that compose us cannot be separated from the space and particles surrounding them, so living beings cannot be isolated from each other. A living system of greater coherence could exchange information and create or restore coherence in a disordered, random or chaotic system. The natural state of the living world appeared to be order – a drive toward greater coherence. Negentropy appeared to be the stronger force. By the act of observation and intention, we have the ability to extend a kind of super-radiance to the world.

This tango appears to extend to our thoughts as well as our bodily processes. Our dreams, as well as our waking hours, may be shared between ourselves and everyone who has ever lived. We carry on an incessant dialogue with The Field, enriching as well as taking from it. Many of humankind's greatest achievements may result from an individual suddenly gaining access to a shared accumulation of information – a collective effort in the Zero Point Field – in what we consider a moment of inspiration. What we call 'genius' may simply be a greater ability to access the Zero Point Field. In that sense, our intelligence, creativity and imagination are not locked in our brains but exist as an interaction with The Field.[34]

The most fundamental question Braud's work raises has to do with individuality. Where does each of us end and where do we begin? If every outcome, each event, was a relationship and thoughts were a communal process, we may need a strong community of good intention to function well in the world. Many other studies have shown that strong community involvement was one of the most important indicators of health.[35]

The most interesting example of this was a small town in Pennsylvania called Roseto. This tiny town was entirely populated with immigrants from the same area of Italy. Along with the people themselves, their culture had been transplanted in its entirety. The town shared a very cohesive sense of community; rich lived cheek by jowl with poor, but such was the sense of interrelation that jealousy seemed to be minimized. Roseto had an amazing health record. Despite the prevalence of a number of high-risk factors in the community – smoking, economic stress, high-fat diets – the people of Roseto had a heart-attack rate less than half that of neighboring towns.

One generation later, the cohesiveness of the town broke up; the youth didn't carry on the sense of community, and before long it began to resemble a typical American town – a collection of isolated individuals. In parallel, the heart-attack rate quickly escalated to that of its neighbors.[36] For those few precious years, Roseto had been coherent.

Braud had shown that human beings trespass over individual boundaries. What he didn't yet know was how far we could travel.

CHAPTER EIGHT

The Extended Eye

DOWN IN THE BASEMENT of a physics building at Stanford University, the tiniest flicker of the tiniest fragments of the world were being captured and measured. The device required to measure the movement of subatomic particles resembled nothing so much as a three-foot hand mixer. The magnetometer was attached to an output device whose frequency is a measure of the rate of change of magnetic field. It oscillated ever so slightly, grinding out its slowly undulating S-curve on an *x–y* recorder, a paper graph, with annoying regularity. To the untrained eye, quarks were sedentary: nothing ever changed on the graph. A non-physicist might look upon this gadget as something akin to a souped-up pendulum.

A Stanford physics student named Arthur Hebard had seen the superconducting differential magnetometer as a fitting post-doctoral occupation, applying for grant money to devise an instrument impervious to all but the flux in the electromagnetic field caused by any quarks which happened to be passing by. Nevertheless, to anyone who understood about measuring quarks, it was a delicate business. It necessitated blocking out virtually all the endless electromagnetic chatter of the universe in order to hear the infinitesimal language of a subatomic particle. To accomplish this, the magnetometer's innards needed to be encased in layer upon layer of shielding – copper shielding, aluminium casing, a superconducting niobium shield, even μ-metal shielding, a metal which specifically limits magnetic field. The device was then buried in a concrete well in the floor of the lab. The SQUID (superconducting quantum interference device) was a bit of a mystery at Stanford – seen but not understood. No one had ever published its complex inner construction.

To Hal Puthoff, the magnetometer was a quackbuster. He looked upon it as the perfect test of whether there was such a thing as psychic power. He was open-minded enough to test whether psychokinesis worked, but not really convinced. Hal had grown up in Ohio and Florida, but liked to say he was from Missouri – the Show Me state, the ultimate state of the skeptic. Show me, prove it to me, let me see how it works. Scientific principles were a comforting refuge for him, the best way he could get a handle on reality. The multiple layers of shielding erected around the magnetometer would present the ultimate challenge for Ingo Swann, the psychic, whose plane was arriving from New York that afternoon. He would spring the thing on Swann. Just let him see if he could alter the pattern of a machine impervious to anything short of an atomic explosion.

It was 1972, the year before he'd begun working on his Zero Point Field theories, when Hal was still at SRI. Even at that time, before he'd thought about the implications of quantum zero-point fluctuations, Hal was interested in the possibility of interconnection between living things. But at this stage, he didn't really have a focus, much less a theory. He'd been dabbling in tachyons, or particles that travel faster than the speed of light. He'd wondered whether tachyons could explain some studies he'd come across showing that animals and plants had the ability to engage in some sort of instantaneous communication, even when separated by hundreds of miles or shielded by a variety of means. Hal had really wanted to find out whether you could use quantum theory to describe life processes. Like Mitchell and Popp, he'd long suspected that everything in the universe on its most basic level had quantum properties, which would mean that there ought to be nonlocal effects between living things. He'd been kicking around an idea that if electrons had nonlocal effects, this might mean something extraordinary on a large scale in the world, particularly in living things – some means of acquiring or receiving information instantaneously. At the time, all he had in mind to test this assumption was a modest study, mainly involving a bit of algae, which Bill Church was eventually persuaded to invest $10,000 in.

Hal had sent the proposal to Cleve Backster, a New York polygraph expert who'd been carrying out studies, just for fun, to see if plants register any 'emotion' – in the form of electrical signaling – on standard lie detector equipment, the same way humans do in response to stress. These were the studies that had so fascinated Hal. Backster tried burning the leaf of a plant and then measured its galvanic response, much as he would register the skin response of a person being tested for lying. Interestingly enough, the plant registered the same increased-stress polygraph response as a human would if his hand had been burned. Even more fascinating, as far as Hal was concerned, was that Backster had burned the leaf of a neighboring plant not connected to the equipment. The original plant, still hooked up to the polygraph, again registered the 'pain' response that it had when its own leaves had been burned. This suggested to Hal that the first plant had received this information via some extrasensory mechanism and was demonstrating empathy. It seemed to point to some sort of interconnectedness between living things.[1]

The 'Backster effect' had also been seen between plants and animals. When brine shrimp in one location died suddenly, this fact seemed to instantly register with plants in another location, as recorded on a standard psychogalvanic response (PGR) instrument. Backster had carried out this type of experiment over several hundred miles and among paramecium, mold cultures and blood samples, and in each instance, some mysterious communication occurred between living things and plants.[2] As in *Star Wars*, each death was registered as a disturbance in The Field.

Hal's proposal for the algae experiments happened to be sitting on Backster's desk the day that he'd been visited by Ingo Swann. Swann, an artist, was mainly known as a gifted psychic, who'd been working on ESP experiments with Gertrude Schmeidler, a professor in psychology at City College in New York.[3] Swann had rifled through Hal's proposal and was intrigued enough to write to him, suggesting that if he were interested in looking at some common ground between the inanimate and the biological that he start doing some experiments in psychic phenomena. Swann himself had done some work on out-of-body experiments and had got good results. Hal was deeply skeptical, but gamely took him up on his suggestion. He contacted Bill Church to see if he could change his study and use some of his grant money to fly Swann out to California for a week.

A short, chubby man with amiable features, Swann arrived dressed absurdly in a white cowboy hat with white jacket and Levis, like some visiting rock star. Hal grew convinced that he was wasting Bill Church's money. Two days after Swann arrived, Hal took him down to the basement of the Varian Hall physics building.

Hal pointed to the magnetometer. He asked Ingo to attempt to alter its magnetic field. Hal explained that any alteration would show up in the output tape.

Ingo initially was disturbed by the prospect, as he'd never done anything like this before. He said he was first going to psychically peer into the innards of the machinery to get a better sense of how to affect it. As he did, the S-curve suddenly doubled its frequency for about 45 seconds – the length of Ingo's time of concentration.

Could he stop the field change on the machine, which is indicated by the S-curve? Hal asked him.

Ingo closed his eyes and concentrated for 45 seconds. For the same length of time the machine's output device stopped creating equidistant hills and valleys: the graph traced one long plateau. Ingo said he was letting go; the machine returned to its normal S-curve. He explained that by looking into the machine and concentrating on various parts, he was able to alter what the machine did. As he spoke, the machine again recorded a double frequency and then a double dip – which Ingo said had something to do with his concentrating on the niobium ball inside the machine.

Hal asked him to stop thinking about it and chatted with him about other subjects for several minutes. The normal S-curve resumed. Now concentrate on the magnetometer, Hal said. The tracing started furiously scribbling. Hal told him to stop thinking about it, and the slow S resumed. Ingo did a quick sketch of what he said he 'saw' as the design of the inside of the machine and then asked if they could stop as he was tired. For the next three hours, the machine's output went back to its regular curves, monotonous and steady.

A group of graduate students who'd gathered around put the changes down to some strange and coincidental electromagnetic noise creeping into the system. As far as they were concerned, a readily explained blip had occurred. But then Hal had the drawing checked out by Hebard, the post-doctoral student who'd created the machine, and he said it was dead-on accurate.

Hal didn't know what to make of it. It appeared that some nonlocal effect had occurred between Ingo Swann and the magnetometer. He went home and wrote a guarded paper on the subject and circulated it to his colleagues, asking them to comment on it. What he'd seen usually went by the name of astral projection or out-of-body experiences, or even clairvoyance, but he would eventually settle on a nice, neutral, non-emotive phrase for it: 'remote viewing'.

Hal's modest experiment launched him on a 13-year project, carried out in parallel with his Zero Point Field work, which sought to determine whether people could see things beyond any known sensory mechanism. Hal realized he'd stumbled on some property of human beings that was not a million miles from what Backster observed – some instant connection with the unseen. Remote viewing seemed of a piece with the notion he'd been toying with about some sort of interconnection between living things. Much later, he would privately speculate about whether remote viewing had anything to do with the Zero Point Field. For the moment, all he was interested in was whether what he'd seen was real and how well it worked. If Swann could see inside magnetometers, was it possible for him to see anywhere else in the world?

Inadvertently, Hal also launched America on the largest spy program ever attempted using clairvoyance. A few weeks after he'd circulated his paper, two blue-suited members of the Central Intelligence Agency arrived at his door, waving the report in hand. The agency, they told him, was getting increasingly concerned about the amount of experiments the Russians were conducting into parapsychology funded by the Soviet security forces.[4] From the resources they were pouring into it, it seemed as though the Russians were convinced that ESP could unlock all of the West's secrets. A person who could see and hear things and events separated by time and space represented the perfect spy. The Defense Intelligence Agency had just circulated a report, 'Controlled offensive behavior – USSR', which predicted that the Soviets, through their psychic research, would be able to discover the contents of top secret documents, the movements of troops and ships, the location of military installations, the thoughts of generals and colonels. They might even be able to kill or shoot down aircraft from a distance.[5] Many senior staff at the CIA thought it was high time that the US looked into it as well; the problem was that they were getting laughed out of most labs. Nobody in the American scientific community would take ESP or clairvoyance seriously. It was the CIA's view that if they didn't, the Russians would probably gain an advantage that the US would never be able to overcome. The agency had been scouring around for a small research lab outside academia that might be willing to carry out a small, low-key investigation. SRI – and Hal's current interest – seemed perfect for the job. Hal even checked out as a good security risk since he'd had experience in intelligence in the Navy and had worked for the National Security Agency.

The men asked Hal to carry out a few simple experiments – nothing elaborate, perhaps just guessing objects hidden in a box. If they were successful, the CIA would agree to fund a pilot program. The two men from Washington later watched Swann correctly describe a moth hidden in the box. The CIA was impressed enough to throw nearly $50,000 at a pilot project, which was to last for eight months.

Hal agreed to continue with the box-guessing exercise and for several months he carried out trials with Ingo Swann, who managed to describe objects hidden in boxes with great precision – far more successfully than could have been achieved by simple guessing.

By that time, Hal had been joined by a colleague in laser physics called Russell Targ, who'd also pioneered development of the laser for Sylvania. It was probably no accident that another physicist interested in the effect of light through space would also be intrigued by the possibility that the mind could also breach vast distances. Like Hal, Targ also checked out as a good security risk for the classified operation because he'd been involved in security studies for Sylvania. Tall and lanky at 6 foot 5, Russ had a shock of curly hair, which sat back on his forehead – a dark-haired Art Garfunkel to Hal's sturdier Paul Simon. There the resemblance ended; anchored to Russ's face was a pair of black Coke-bottle glasses. Targ had terrible vision and was considered legally blind. Even his glasses only corrected his sight to a fraction of normal. His poor outward vision may have been one reason why he saw pictures in his mind's eye so clearly.

Targ had become interested in the nature of human consciousness from his hobby as an amateur magician. Many times up on the stage, he'd be performing some conjuring trick about his subject, taken from the audience, and although he'd have rigged the actual trick, he'd suddenly realize in the midst of it that he knew more information than he'd been told. He might be pretending to guess a question about a location and suddenly a clear mental image of it would pop into his head. Invariably, his own internal picture would turn out to be accurate, which only enhanced his reputation as a magician, but left him with many questions about how this could possibly be happening.

It had been Ingo's idea to try his hand at a real test of his powers – one that would more closely resemble how the CIA figured remote viewing ought to be used. He had the idea of using geographical coordinates as a quick, clean, non-emotive way to get to the spot. Both Puthoff and Targ were skeptical of such an idea. If they gave him coordinates and Swann guessed correctly, it might simply mean that he'd remembered a site on a map – he might have a photographic memory.

They made a few desultory attempts, and Swann was way off target. But then, after fifty attempts, Swann began to improve. By Swann's 100th coordinate, Hal was impressed enough to get on the phone to Christopher Green, an analyst in the CIA's Office of Scientific Intelligence, urging him to allow them to try a real test for the agency. Although Green was highly dubious, he agreed to give them a set of map coordinates of a place not even he knew anything about.

A few hours later, at Green's request, a colleague named Hank Turner[6] produced a set of numbers on a sheet of paper. These represented extremely precise coordinates, down to the minutes and seconds of latitude and longitude, of a place that only Turner knew. Green took the paper and picked up the phone to call Hal.

Puthoff sat Swann down at a table at SRI and gave him the coordinates. As he puffed on a cigar, and alternated between closing his eyes and scribbling on a piece of paper, Swann described a burst of images: 'mounds and rolling hills', 'a river over to the far east', 'a city to the north'. He said it seemed to be a strange place, 'somewhat like the lawns that one would find around a military base'. He got the impression that there were 'old bunkers around', or it could simply be 'a covered reservoir'.[7]

The following day, Swann tried again at home, and jotted down his impressions on a report which he'd brought in to Hal. Again, he got the impression that something was underground.

A few days later, Puthoff received a phone call from Pat Price, a building contractor from Lake Tahoe, who also raised Christmas trees. Price, who considered himself a psychic, had met Puthoff at a lecture and was calling now to offer his services in their experiments. A florid, wisecracking Irishman in his early fifties, Price said he'd been using his own version of remote viewing successfully for many years, even to catch criminals. He'd served briefly as police commissioner in Burbank, a suburb of Los Angeles. Price would be in the dispatch room and as soon as a crime had been reported, he'd scan the city mentally. Once he settled on a place, he'd immediately send a car to the location in his mind. Invariably, he claimed, he'd caught his man, just at the spot he'd visualized.

On a whim, Puthoff gave Price the coordinates given to him by the CIA. Three days later, Hal received a package Price had posted the day after they'd spoken, containing pages of descriptions and sketches. It was obvious to Puthoff that Price was describing the same place as Swann, but in far more detail. He offered a highly precise description of the mountains, the location of the place, and its proximity to roads and a town. He even described the weather. But it was the interior of one peak area that interested Price. He wrote that he thought he saw an 'underground storage area' of some variety which had been well concealed, perhaps 'deliberately so'.

'Looks like former missile site – bases for launchers still there, but area now houses record storage area, microfilm, file cabinets,' he wrote. He was able to describe the aluminum sliding doors, the size of the rooms and what they contained, even the large maps pinned on the wall.

Puthoff phoned Price and asked him to look again, to pick up any specific information, such as code names or the names of officers. He wanted to take this to Green and needed details to dispel any lingering disbelief. Price returned with details from one specific office: files named 'Flytrap' and 'Minerva', the names on labels on folders inside filing cabinets, the names of the colonel and majors who sat at the steel desks.

Green brought the information to Turner. Turner read their reports and shook his head. The psychics were totally off beam, he said. All he'd given him were the coordinates of the location of his summer cabin.

Green went away, puzzled by the fact that both Swann and Price had described so similar a place. That weekend, he drove out to the site with his wife. A few miles from the coordinates, down a dirt road, he found a government 'No Trespassing' sign. The site seemed to match the descriptions of both psychics.

Green began inquiring about the site. Immediately he got embroiled in a heated investigation of a security breach. What Swann and Price had correctly described was a vast secret Pentagon underground facility in the Blue Ridge Mountains of West Virginia, manned by National Security Agency code breakers, whose main job was to intercept international telephone communications and control US spy satellites. It was as though their psychic antennae had picked up nothing of note with the original coordinates and so scanned the area until they got on the wavelength of something more relevant to the military.

For months, the NSA was convinced that Puthoff and Targ, and even Green himself, were being provided this information from some source within the facility. Puthoff and Targ were checked out as security risks and their friends and associates questioned as to their communist leanings. Price only managed to calm down the agency by throwing it a bone: detailed information about the Russian counterpart to the NSA's secret site, operated by the Soviets in the northern Ural Mountains.

After the West Virginia episode, CIA officials at the highest levels were convinced enough to try a real test in the field. One day, one of the contract monitors came to SRI with the geographical coordinates of a Soviet site of great concern to the agency. All Russ and Hal were told was that the site was an R&D test facility.[8]

Price was the one they wanted to test. Targ and Price headed up to the special room, housed on the second-floor of the Radio Physics building – which had been electrically shielded with a double-walled copper screen, which would block a remote viewer's ability if it were generated by a high-frequency electromagnetic field. Targ started the tape. Pat removed his wire-rim glasses, leaned back in his chair, took a crisp white linen handkerchief from his pocket, polished his glasses, then closed his eyes, and only spoke after a full minute.

'I am lying on my back on the roof of a two- or three-storey brick building,' he said dreamily. 'It's a sunny day. The sun feels good. There's the most amazing thing. There's a giant gantry crane moving back and forth over my head . . . As I drift up in the air and look down, it seems to be riding on a track with one rail on each side of the building. I've never seen anything like that.'[9] Pat went on to sketch the building layout and paid particular attention to what he kept describing as a 'gantry crane'.

After two or three days, once they'd finished the work on that site, Russ, Hal and Pat were astonished to hear that they'd had been asked about a suspected PNUTS, which is CIA-code for a Possible Nuclear Underground Testing Site. This place was driving the agency crazy. Everything in America's intelligence arsenal was being thrown at this spot, to find out what on earth was going on inside. Pat's drawing turned out to be extremely close to satellite photos, even down to a cluster of compressed-gas cylinders.

Pat didn't stop at the outside of the building. His descriptions included what was going on inside. He saw images of workers attempting, with great difficulty, to assemble a massive 60-foot metal globe by welding together metal gores, shaped like wedges of fruit. However, the pieces were warping and Pat believed they were attempting to find material they could weld at lower temperatures.

No one in the government had any idea of what was going on inside the facility and Pat died a year later. Nevertheless, two years later, an Air Force report was leaked to *Aviation Week* magazine about the CIA's use of high-resolution photographic reconnaissance satellites, which finally confirmed Pat's vision. The satellites were being used to observe the Soviets digging though solid granite formations. They'd been able to observe enormous steel gores being manufactured in a nearby building.

'These steel segments were parts of a large sphere estimated to be about 18 meters (57.8 feet) in diameter', said the *Aviation Week* article.

'US officials believe that the spheres are needed to capture and store energy from nuclear driven explosives or pulse power generators. Initially, some US physicists believed that there was no method the Soviets could use to weld together the steel gores of the spheres to provide a vessel strong enough to withstand pressures likely to occur in a nuclear explosive fission process, especially when the steel to be welded was extremely thick.'[10]

When Pat's drawings matched the satellite photos so well, the CIA assumed the nuclear spheres he saw must be manufactured for atomic bombs, and one assumption after another led the Reagan Administration to dream up what became known as the Star Wars program.[11] Many billions of dollars later, it turned out to be a curve ball. Semipalatinsk, the site Pat had seen, wasn't even a military installation. The Russians indeed were trying to develop nuclear rockets, but for their own manned Mars mission. All the rockets were to be used for was fuel.

Pat Price couldn't tell the American government what Semipalatinsk was used for, and he died before he could warn them off Star Wars. But for Targ and Puthoff, the Semipalatinsk sighting meant more than just a bit of psychic spying. This gave them some vital evidence about how remote viewing worked. Here was evidence of an individual who could take geographical coordinates anywhere in the world and directly see and experience what was going on there, even at a site that no one in the US had any knowledge of.

But was any distance too far? The other amazing experiment was conducted with Ingo Swann. Swann was also interested in testing their assumption that a human beacon needed to be present at a site for a remote viewer to pick it up. He had a bold suggestion – a test that might strain all his skills. Why didn't he try to view the planet Jupiter, just before the upcoming NASA *Pioneer 10* flyby launch?

During the experiment, Swann was embarrassed to admit that he'd seen – and drawn – a ring around Jupiter. Perhaps, he told Puthoff, he'd just mistakenly directed his attention toward Saturn. No one was prepared to take the drawing seriously, until the NASA mission revealed that Jupiter indeed had a ring at the time.[12]

Swann's experiment demonstrated that no individual needed to be present and also that humans could, in effect, 'see' or gain access to information at virtually any distance – something that Ed Mitchell had also found with his card tests when traveling to and from the moon.

Puthoff and Targ wanted to create a scientific protocol for remote viewing. Gradually they moved away from coordinates to places. They created a box file which contained 100 target sites – buildings, roads, bridges, landmarks – within half an hour of SRI, from the San Francisco Bay area to San Jose. All were sealed and prepared by an independent experimenter and locked in a secure safe. An electronic calculator programmed to choose numbers randomly would be used to select one of the target locations.

On the day of the experiment, they'd closet Swann or Price in the special room. One of the experimenters, usually Targ, because of his bad eyesight, would remain behind with Swann. Meanwhile, Hal and one of the other program coordinators would pick up the sealed envelope and head off to the target location, which was not disclosed to either the volunteer or Targ. Hal acted as the 'beacon' of focus – they'd wanted to use someone familiar to Swann or Price whom they could tune in on when attempting to find a mundane location. At the agreed start time, and for the next 15 minutes, Swann was asked to attempt to draw and describe into a tape recorder any impressions of the target site. Targ also would be ignorant of the location of the target team, so that he'd be free to ask questions without fear of inadvertently cueing Swann on the right answer. As soon as the target team returned, they would take the remote viewer to the target site, so that he'd get direct feedback of the accuracy of what he thought he'd seen. Swann's track record was astonishing. In test after test, he had a high accuracy in correctly identifying his target.[13]

With time, Price took over as chief remote viewer. Hal and Russ underwent nine trials with him, following their usual double-blind protocol of sealed target spots near Palo Alto – Hoover Tower, a nature preserve, a radio telescope, a marina, a toll plaza, a drive-in movie theater, an arts and crafts plaza, a Catholic church and a swimming pool complex. Independent judges concluded that Price had scored seven hits out of the nine. In some cases, like the Hoover Tower, Price even recognized it and correctly identified it by name.[14] Price was noted for his incredible accuracy and also his ability to 'see' through the eyes of his traveling partner. One day, when Puthoff traveled to a boat marina, Pat shut his eyes, and when he opened them, blurted out, 'What I'm looking at is a little boat jetty or boat dock along the bay . . .'[15]

Hal even tested Pat on detail. He sent Green, the CIA boss, up in a small aircraft with three numbers on a piece of paper inside his breast pocket. Numbers and letters were known to be almost impossible to remote view accurately. Nevertheless, there was Pat Price ticking them off, even in order. He only complained of feeling a bit seasick and drew a picture of a kind of special cross, which he'd had the image of swinging back and forth, making him ill. It turned out that Green was wearing an *ankh*, an ancient Egyptian cross matching Price's drawing, around his neck, and the necklace must have been swinging wildly during the ride.[16]

Although the results of Price and Swann had been impressive, the Agency wanted to convince itself that this was not simply the work of the highly gifted or, even worse, an elaborate conjuring trick. A couple of the CIA contract monitors asked if they could try their hand at it. This appealed to Hal, who'd wanted to see whether ordinary individuals could carry out remote viewing. Each was invited to participate in three experiments, and both improved with practice. The first scientist correctly identified a child's merry-go-round and a bridge, and the second correctly picked up a windmill. Of the five experiments, three were direct hits and one a near miss.[17]

When the CIA's test studies worked, Puthoff and Targ began gathering up ordinary volunteers, some naturally gifted, but unpracticed in remote viewing, some not. In late 1973 and early 1974, Puthoff and Targ selected four ordinary people, three of them SRI employees and one a photographer named Hella Hammid, a friend of Targ's. Hammid, who'd never been involved in psychic research before, turned out to be a natural at remote viewing. In five of nine targets, Hella scored direct hits, as determined by independent judges.[18]

Hal needed to go to Costa Rica for business, so he decided to use the trip to act as a long-distance target. On each day of his trip, he would keep a detailed record of his location and activities at precisely 1:30 p.m. Pacific daylight time. At the same time, Hella or Pat Price would be asked to describe and draw where Dr Puthoff was every day at that time.

One day, when neither Hella nor Pat showed up, Targ stood in their place as the remote viewer. He got a strong sense that Puthoff was at an ocean or beach setting, even though he knew that Costa Rica is primarily a mountainous country. Although dubious about his accuracy, he described an airport and airstrip on a sandy beach with an ocean at one end. At that moment, Hal had taken an unplanned diversion to an offshore island. At the designated time, he was just getting out of a plane at a tiny island airport. In every regard, save one, Targ described and drew the airport accurately. The only small error had to do with his drawing of the airport; he'd drawn a building looking like a Quonset hut, when in fact the building was rectangular. During the rest of his trip, Hammid and Price correctly identified when Hal was relaxing round a pool or driving through a tropical forest at the base of a volcano. They were even able to identify the color of his hotel rug.[19]

Hal gathered together nine remote viewers in total, mostly beginners with no track record as psychics, who performed in total over fifty trials. Again, an impartial panel of judges compared targets with transcripts of subject descriptions. The descriptions may have contained some inaccuracies, but they were detailed and accurate enough to enable the judges to directly match description with target roughly half the time – a highly significant result.

As a backup method of judging the accuracy of the viewing, Hal then asked a panel of five SRI scientists not associated with the project to blind-match unedited, unlabeled typed transcripts and drawings made by the remote viewers with the nine target sites, which they visited in turn. Between them, the judges came up with twenty-four correct matches of transcript with target site, against an expected five.[20]

By degrees, Puthoff and Targ were turning into believers. Human beings, talented or otherwise, appeared to have a latent ability to see anywhere across any distance. The most talented remote viewers clearly could enter some framework of consciousness, allowing them to observe scenes anywhere in the world. But the inescapable conclusion of their experiments was that anyone had the ability to do this, if they were just primed for it – even those highly skeptical of the entire notion. The most important ingredient appeared to be a relaxed, even playful, atmosphere which deliberately avoided causing anxiety or nervous anticipation in the viewer. And that was all, other than a little practice. Swann himself had learned over time how to separate signal from noise – somehow divining what was his imagination from what was clearly in the scene.

Puthoff and Targ had tackled remote viewing as scientists, creating a scientific method for testing it. Brenda Dunne and Robert Jahn refined this science even further. This was a natural progression for them. One of the first to replicate the SRI work had been Brenda Dunne, while an undergraduate at Mundelein College and later as a graduate student of the University of Chicago, before her move to Princeton.[21] Dunne's forte, once again, had been ordinary volunteers, not gifted psychics. In eight studies using two students with no gift for psychic ability, she demonstrated that her participants could be successful in correctly describing target locations. Once she joined Princeton, remote viewing also became included in PEAR's agenda.

Jahn and Dunne were mainly worried about the great likelihood that these sorts of studies would be vulnerable to sloppy protocols and data-processing techniques or deliberate or inadvertent 'sensory cueing' by either participant. Determined to avoid any of these weaknesses, they were painstaking in study design. They came up with the latest subjective way of measuring success – a standardized checklist. Besides describing the scene and drawing a picture, the remote viewer would be asked to fill in a form of thirty multiple-choice questions about the details of the scene, which attempted to give flesh to the bones of his or her description. Meanwhile, the person at the remote site would also fill in the same form, in addition to taking photos and making drawings. On many occasions, the target site was selected by one of the REG machines and handed in a sealed envelope to the traveler, to be opened away from PEAR; on other occasions, the traveling participant might choose a target site only after he or she was at a remote site unknown to anybody back at Princeton.

When the traveler returned, a member of the PEAR staff would enter the data into a computer, which would compare checklists for the traveler and remote viewer, and also compare these lists with all others in the database.

In total, Jahn and Dunne performed 336 formal trials involving 48 recipients and distances between traveler and remote viewer of between 5 and 6000 miles, and worked out a highly detailed mathematical analytical assessment to judge the accuracy of the results. They even determined individual probability scores for arriving at the right answer by chance. *Nearly two-thirds were more accurate than could be accounted for by chance.* The overall odds against chance in the PEAR's complete remote viewing database was one billion to one.[22]

One possible criticism was that most of the remote viewing pairs knew each other. Although some sort of emotional or physiological bond between the participants seemed to improve the scores, good results were also achieved when the traveler and remote viewer were virtual strangers. Unlike the initial SRI studies, no one was chosen because of a gift for telepathy. Furthermore, better scores were obtained when the traveling participants were randomly assigned their sites from a large pool of possibilities, rather than spontaneously selecting it themselves. This made it unlikely that any common knowledge between the pairs of participants improved the scores.

Jahn, as well as Puthoff, realized that nothing in the current theories of biology or physics could account for remote viewing. The Russians had maintained that clairvoyance operated through some sort of extremely-low-frequency (ELF) electromagnetic wave.[23] The problem with this interpretation is that in many of the experiments, the viewers had been able to see a site as a moving video, as if they had been there on the scene. This meant that this phenomenon operated beyond a conventional ELF frequency. Furthermore, using the special double-walled, copper-screened room, which would block even low-frequency radio waves, didn't tarnish anyone's ability to pick up the scene or degrade any of the descriptions, even those of events thousands of miles away.

Puthoff went on to test the ELF hypothesis by conducting two of their studies from a Taurus submarine, a tiny five-person vehicle made by the International Hydrodynamics Company Ltd (HYCO) of Canada. Several hundred feet of sea water is known to be an effective shield for all but the very lowest frequencies of the electromagnetic spectrum. The remote viewer – usually Hammid or Price – traveled in the submarine 170 metres under the surface near Catalina Island, off the coast of Southern California, while Hal and a government contract monitor picked out a target from a pool of target locations near San Francisco. At the designated time, they went to the site and stayed for 15 minutes. At this point, Hammid or Price would try to describe and draw what her or his partner was looking at 500 miles away.

In both cases, they'd correctly identified the target site – a tree on a hilltop in Portola Valley and a shopping mall in Mountain View. This made it highly unlikely that the channel of communication was electromagnetic waves, even of extremely low frequency. Even the very low 10 Hz brain waves would be blocked in 170 metres of water. The only waves that wouldn't be blocked were quantum effects. As every object absorbs and re-radiates the Zero Point Field, the information would be re-emitted back through the other side of the water 'shield'.

Puthoff and Targ did have a few clues about the peculiar characteristics of remote viewing. For one thing, each of the SRI remote viewers appeared to have his or her own signature. Orientation appeared to match a person's tendencies in other regards; a sensory remote viewer would also view with his or her senses in person. One might be particularly good at mapping out the site and describing the architectural and topographical features; another would concentrate on the sensory 'feel' of the target; yet another would focus on the behavior of the target experimenter, or describe what he was feeling and seeing, as though he was somehow transported and able to see out of the target person's eyes.[24] Many of the viewers operated in 'real time' as though they were somehow there, experiencing the scene from their target subject's point of view. When Hal was swimming in Costa Rica, they saw the scene from his perspective; if he was distracted by a scene other than the central one he was visiting at the time, then so were they. It was as though they operated with the senses of two people – their own and the person on the scene.

The signals were acting as though they'd been sent through some low-frequency bit channel. The information in their experiments was received in bits and often imperfectly. Although the basic information came through, sometimes the details were a little blurred. Usually, the scene was flipflopped, so that the subject would see the reverse, as though looking at the scene through a mirror. Targ and Puthoff had wondered whether this might have to do with the ordinary activity of the visual cortex, as they understood it. The conventional view was that the cortex takes in a scene in reverse, and the brain corrects this by switching the scene. In this instance, the sight isn't being viewed by the eyes, but the brain still performs its reverse correction of the scene. That is where the similarity with ordinary brain activity ended. Many of the remote viewers had been able to change their perspective, particularly when gently urged to do so by their monitor, so they could move around heights and angles at will, or zoom in for a close up, like a video camera on a crane. With Pat's first remote viewing of the secret Pentagon site, he'd begun his viewing from 1500 feet up to take the scene in as a whole and then zoomed in for closer detail.

The worst thing a remote viewer could do was to interpret or analyze what he saw. This tended to color his impressions as the information was still filtering through, and invariably, he would guess wrong. Based on that guess, he would begin to interpret other items in the scene as being likely companions to the interpreted main image. If one viewer thought he saw a castle, he'd begin looking for a moat. His expectation or imagination would take the place of the receiving end of the channel.[25] There was no doubt that information came through spatially and holistically in flashes of images. As with the phenomena studied by PEAR and Braud, this sensory channel appears to make use of the unconscious and nonanalytic part of the brain. As Dunne and Jahn had found with their REG machines, the left brain is the enemy of The Field.

Remote viewers were exhausted when they finished and also overwhelmed by a kind of sensory overload when they returned to the here and now. It was as though they'd entered into some super consciousness, and once they'd come out of it, the world was more intense. The sky was bluer, sounds were louder, everything more deliciously real. It was as if, in tuning in to those barely perceptible signals, their senses had been turned to maximum. Once they rejoined the world, ordinary volume bombarded them with sight and sound.[26]

Hal began to think about how remote viewing might be possible. He didn't want to attempt a theory. Like most scientists, he hated woolly speculation. But there was no doubt that at some level of awareness, we had all information about everything in the world. Clearly, human beacons weren't always necessary. Even a set of coordinates could take us there. If we could see remote places instantaneously, it argued strongly that it was a quantum, nonlocal effect. With practice, people could enlarge their brain's receiving mechanisms to gain access to information stored in the Zero Point Field. This giant cryptogram, continually encoded with every atom in the universe, held all the information of the world – every sight and sound and smell. When remote viewers were 'seeing' a particular scene, their minds weren't actually somehow transported to the scene. What they were seeing was the information that their traveler had encoded in quantum fluctuation. They were picking up information contained in The Field. In a sense, The Field allowed us to hold the whole of the universe inside us. Those good at remote viewing weren't seeing anything invisible to all the rest of us. All they were doing was dampening down the other distractions.

As every quantum particle is recording the world in waves, carrying images of the world at every moment, at some profoundly deep quantum level, something about the scene – a target individual or map coordinates – is probably acting like a beacon. A remote viewer picks up signals from the target individual and the signal carries an image that is picked up by us at a quantum level. To all but the experienced and the gifted, like Pat Price, this information is received imperfectly, in reverse or in incomplete images, as if something were wrong with the transmitter. Because the information is received by our unconscious mind, we often receive it as we would in a dream state, a memory or a sudden insight – a flash of an image, a portion of the whole. Price's success with the Russian site and Swann's success with Jupiter suggest that any sort of mnemonic, such as a map or cipher, can conjure up the actual place. As an idiot savant has access to impossible calculations in an instant, perhaps the Zero Point Field enables us to hold an image of the physical universe inside ourselves, and under certain circumstances we open our bandwidths wide enough to glimpse a portion of it.

The SRI remote viewing program (later housed at the Science Applications International Corp, or SAIC) carried on for twenty-three years, behind a wall of secrecy that is still erected. It had been funded entirely by the government, first under Puthoff, then Targ and finally Edwin May, a burly nuclear physicist who'd carried out other intelligence work before. In 1978, the Army had its own psychic spying intelligence unit in place, code-named Grill Flame, possibly the most secret program in the Pentagon, manned by enlisted men who'd claimed some talent in psychic phenomena. By the time of Ed May's tenure, a who's who of scientists consisting of two Nobel laureates and two chairs of department at universities, all chosen for their skepticism, sat on a government Human Use and Procedural Oversight committee. Their task was to review all of the SRI remote viewing research, and to do so they were given unannounced drop-in privileges to SAIC, to guard against fraud. All concluded that the research was impeccable, and half actually felt the research demonstrated something important.[27] Nevertheless, to this day, the American government has released only the Semipalatinsk study, one tiny portion of a mountain of SRI documents, and then only after a relentless campaign by Russell Targ.[28]

At the close of the program in 1995, a government-sponsored review of all the SRI and SAIC data, carried out by Jessica Utts, a statistics professor at the University of California at Davis, and Dr Ray Hyman, a skeptic of psychic phenomena, agreed that the statistical results for remote viewing phenomena were far beyond what could have occurred by chance.[29] As far as the US government was concerned, the SRI studies gave America a possible advantage over Russian intelligence. But to the scientists themselves, these results represented far more than a chess maneuver in the Cold War. It seemed to suggest that because of our constant dialogue with the Zero Point Field, like de Broglie's electron, we are everywhere at once.

CHAPTER NINE

The Endless Here and Now

THE CIA MIGHT HAVE been struck by Pat Price's success with Semipalatinsk, but that wasn't the experiment which most impressed Hal Puthoff and Russell Targ. That one had occurred the year before and concerned nothing more cloak and dagger than a local swimming pool.

Targ had been with Pat Price in the copper-screened room on the second floor of the SRI Radio Physics building; Hal and a colleague had their electronic calculator randomly choose one of the locations, which in this instance turned out to be the swimming-pool complex in Rinconada Park in Palo Alto, approximately five miles away.

After 30 minutes, when it was likely that Puthoff had arrived at his destination, Targ gave Price the go-ahead. Price closed his eyes and described in detail, and with near-correct dimensions the large pool, the smaller pool and a concrete building. In all respects his drawing was accurate, save one: he insisted that the site housed some sort of water purification plant. He even drew rotating devices into his drawings of the pools and added two water tanks on site.

For several years, Hal and Russell had just assumed that Pat had got this one wrong. Too much noise to signal is how they usually phrased it. There was no water purification system there, and there certainly weren't any water tanks.

Then, in early 1975, Russell received an Annual Report of the City of Palo Alto, a celebration of its centennial, containing some of the city's highlights over the last century. While flicking through it, Targ was flabbergasted to read: 'In 1913 a new municipal waterworks was built on the site of the present Rinconada Park.' It also included a photo of the site, which clearly showed two tanks. Russ remembered Pat's drawing and pulled it out; the tanks were exactly in the place that Pat Price had drawn them. When Pat 'saw' the site, he saw it as it had been 50 years ago, even though all evidence of the water purification plant had long since disappeared.[1]

One of the most astonishing aspects of the data amassed by Puthoff, Jahn and the other scientists is that they hadn't been at all sensitive to distance. A person doesn't have to be in close proximity to affect a REG machine. In at least a quarter of Jahn's studies, the participants were anywhere from next door to thousands of miles away. Nevertheless, the results were virtually identical to those obtained when the participants were at the PEAR lab, sitting right in front of a machine. Distance, even great distance, didn't seem to lessen a person's effect on the machine.[2]

The same had occurred with PEAR's and SRI's remote viewing studies. Remote viewers were able to see across countries, over continents – even out into space.[3]

But the Pat Price study was an example of something even more extraordinary. The research that was emerging from labs such as PEAR and SRI suggested that people could 'see' into the future or reach back into the past.

One of the most inviolate notions in our sense of ourselves and our world is the notion of time and space. We view life as a progression that we can measure through clocks, calendars and the major milestones of our lives. We are born, we grow up, we get married and have children, and one by one collect houses, possessions, cats and dogs, all the while inevitably getting older and moving in a line toward death. Indeed, the most tangible evidence of the progression of time is the physical fact of our own ageing.

The other inviolate notion from classical physics is the notion that the world is a geometric place filled with solid objects with spaces in between them. The size of the space in between determined the kind of influence one object had on another. Things couldn't have any kind of instantaneous influence if they happened to be miles away.

The Pat Price studies and the PEAR studies began to suggest that at a more fundamental level of existence, there is no space or time, no obvious cause and effect – of something hitting something else and causing an event over time or space. Newtonian ideas of an absolute time and space or even Einstein's view of a relative space-time are replaced by a truer picture – that the universe exists in some vast 'here' where here represents all points of space and time at a single instant. If subatomic particles can interact across all space and time, then so might the larger matter they compose. In the quantum world of The Field, a subatomic world of pure potential, life exists as one enormous present. 'Take time out of it,' Robert Jahn was fond of saying, 'and it all makes sense.'

Jahn had his own store of evidence showing that people could foretell events. Largely because of similar work conducted by Brenda Dunne at

Mundelein College, Dunne and Jahn had designed most of their remote viewing studies as 'precognitive remote perception', or PRP. The remote viewers remaining behind in the PEAR lab were asked to name their traveling partner's destinations not only before they actually got there, but also many hours or days before they even knew where they were going. Someone not involved in the experiment would use a REG to randomly pick the traveler's destinations from a pool of previously chosen targets, or the traveler could choose the destination spontaneously and on his own, after setting off. The traveling partner would then follow the standard protocol of remote viewing experiments. He'd spend 10 to 15 minutes at the target site, at the assigned time, recording his impressions of it, taking photos and following the checklist of questions produced by the PEAR team. Meanwhile, back at the laboratory, the remote viewer would have to record and draw his or her impressions of the traveler's destination, *from half an hour to five days before the traveler arrived.*

Of PEAR's 336 formal trials involving remote viewing, the majority were set up as PRP or 'retrocognition' – hours or days after the traveler had left his destination – and were just as successful as those carried out in 'real time'.

Many of the recipients' descriptions matched the traveler's photographs with breathtaking accuracy. In one case, the traveler headed to the Northwest Railroad Station in Glencoe, Illinois, and took one photo of the station with an oncoming train and then another of the inside of the station, a drab little waiting room with a bulletin board below a sign. 'I see the train station,' wrote the remote viewer 35 minutes before the traveler had even chosen where he was going, 'one of the commuter train stations that's on the expressway – the white cement of them and the silver railings. I see a train coming . . . I see or hear the clicking of feet or shoes on the wooden floor. . . . There are posters or something up, some kinds of advertisements or posters on the wall in the train station. I see the benches. Getting the image of a sign . . .'

In another instance, the remote viewer at the PEAR lab jotted down his 'strange yet persistent' image that the agent was standing inside a 'large bowl' – and 'if it was full of soup [the agent] would be the size of a large dumpling'. Forty-five minutes later, the traveler was indeed the size of a dumpling in comparison to the massive curved dome-like structure of the radio telescope in Kitt Peak, Arizona, he was standing under. Yet another PEAR participant described his partner in a 'old building' with 'windows like arches' which 'come to a point on top almost' but 'not a regular point', plus 'great big double doors' and 'square pillars with balls on top'. Nearly a day later, the traveler arrived at his destination, the Tretiakovskaia Gallereia in Moscow, an ornate impressive building with special pillars in front, and a large double door beneath a pointed archway.[4]

In other cases, the remote viewer picked up an impression of a scene on the traveler's journey other than the 'official' one. On one occasion, the traveler intended to visit the Saturn moon rocket at the NASA Space Center in Houston, Texas. The remote viewer, meanwhile, 'saw' an indoor scene where the traveler was playing on the floor with a group of puppies. But that same evening, the traveler (who knew nothing of the remote viewer's impressions) visited a friend's home, where he did indeed play with a litter of newborn puppies, one of which he was prompted to take home with him.

The remote viewers even picked up information about events or scenes that had distracted their travelers from their main targets. One traveler, standing on a farm in Idaho and concentrating on a herd of cows, was distracted by an irrigation ditch several yards down the road. He was sufficiently fascinated by the ditch to photograph it and note it in his description. The remote viewer in New Jersey, picking up the scene before it had happened, made no mention of cows at all in his description, but he did say that he was getting an image of farm buildings, fields and the irrigation ditch.[5]

Other scientific evidence supported the idea that human beings have the ability to 'see' the future. The Maimonides Center's Charles Honorton put together a review of all well-conducted scientific experiments of most varieties. Usually they entailed having participants guess which lamps would light, what card symbols would be turned up, what number on a set of dice would be thrown or even what the weather might be.[6] Combining a total of 2 million trials comprising 309 studies and 50,000 participants, where the time between guessing and the event ranged from a few milliseconds to an entire year, Honorton found positive results with odds against them occurring by chance of ten million billion billion to one.[7]

President Abraham Lincoln dreamed about his own assassination a week before he died. This is one of many good stories about premonitions and dreams foretelling the future that have entered into history. The problem for most scientists is how to test stories like this in the laboratory. How do you quantify and control for a premonition?

The Maimonides dream laboratory had attempted just this – to reproduce people's dreams about their own futures in a credible scientific experiment. They'd come up with a novel procedure, using a gifted English psychic called Malcolm Bessent. Bessent had honed his special talent, studying many years at the London College of Psychic Studies under equally gifted and experienced hands in ESP and clairvoyance. Bessent was invited to sleep at the Maimonides laboratory, where he was asked to dream about what would happen to him the following day. During the night, he would be awakened and asked to report and record his dreams. In one instance, Bessent had followed the agreed procedure for reporting his dream. The next morning, another investigator who'd had no knowledge or contact with Bessent or his dream carried out the agreed procedure for randomly selecting a target among some art reproductions of paintings. It turned out to be Van Gogh's Hospital Corridor at Saint-Remy. As a further precaution against bias, the tape of Bessent's recounting of his dream had been wrapped up and mailed to a transcriber before the picture had been chosen.

As soon as the image was chosen, the Maimonides staff went into high gear. When Bessent woke up and left the sleep room, he was greeted by staff in white coats, who called him 'Mr Van Gogh' and treated him in a rough, perfunctory manner. As he walked along the corridor he could hear the sound of hysterical laughter. The 'doctors' forced him to take a pill and 'disinfected' him with a swab of cotton.

Later, the transcript of his description of his dream was examined. It turned out that Bessent had described a patient attempting to escape, while many people dressed in white coats – doctors and other medical staff – were hostile to him.[8]

Bessent's laboratory premonitions had been highly successful, with seven of a total of eight considered right on target. In a second series, Bessent proved he was able to successfully dream about future targets as well as those he'd just seen. By the time the dream lab was closed in 1978 through lack of funding, they'd amassed 379 trials, with an astonishing 83.5 per cent success rate of present and future dreams.[9]

Dean Radin thought of a novel twist for how to test for a premonition. Instead of relying on verbal accuracy, he'd test whether our bodies were registering any foreboding of an event. This idea was a simplified variation on the dream research. The Maimonides tests were expensive, requiring eight to ten people and a day or so for each experiment. With Radin's protocol, you could get the same results in 20 minutes, at a fraction of the cost.

Radin was part of the small inner circle of consciousness investigators, and one of the only scientists who'd deliberately chosen this field of investigation rather than coming to it through the back door. His involvement in this particular brand of research had to do with the peculiar marriage his life had made of science and science fiction. Radin was 50, but despite the presence of a thin black moustache and a receding hairline, he'd retained the knowing, childlike look of the child prodigy he'd once been. His particular instrument of precocity had been the violin, which he'd played from the age of five up until his mid-twenties. Only lack of physical stamina had caused him give up what might have been a promising career as a concert violinist. World-class musical performance requires nothing less than a superb athlete willing to practice and play for hours every day, honing the mechanics of fine motor control, and Radin came to realize that nothing in his spare physical makeup possessed that level of robustness. It was natural that he would move on to his next great love, fairy tales – the prospect of a secret, magical world. But the same type of precision and detachment that had led to his competence with the violin also made for a skillful investigator, a natural for studying forensic evidence or digging out elusive clues. His first-grade teacher noted the matter-of-fact forthrightness and seriousness of purpose in this slight child and correctly forecast his future vocation. What Radin really wanted to bring into his own juvenile laboratory was magic. He'd wanted to take magic apart and study it under a microscope. By the age of twelve he'd already begun carrying out his own ESP investigations.

Through ten years of university schooling, first in engineering, then a doctorate in psychology, and even a first job in the human factors division of Bell Laboratories, the workings of consciousness and the outer limits of human potential continued to be his chief passion. He'd heard of Helmut Schmidt's machines, and before long he paid Schmidt a visit and came away with a borrowed RNG to conduct some studies of his own. Almost immediately, Radin began getting good results – results as good as Schmidt's. This was too important to be a career sideline. Radin lobbied to work with some of the scientists already in this field, and began doing the rounds, at one point working at SRI and then at Princeton University before setting up his own consciousness laboratory at the University of Nevada in Las Vegas, a remote academic outpost where he hoped he might be left alone.[10]

Radin's initial contribution to this research was the hard statistical grind. Much of his earlier work entailed replicating or providing mathematical verification of the research of his colleagues. It was he who'd worked out the meta-analysis of the PEAR REG studies, among others.

Radin had studied the dream-research data that existed on premonitions. What interested him was whether people had the same sort of clear foreboding when they were awake. In his lab in Las Vegas, Dean set up a computer that would randomly select photos designed either to calm or to agitate, arouse or upset the participant. Radin's volunteers would be wired up to physiological monitors that recorded changes in skin conductance, heart rate and blood pressure.

The computer would randomly display color photos of tranquil scenes (pictures of nature or landscapes) or scenes designed to shock or to arouse (pictures from autopsies or erotic materials). As expected, the participant's body would calm down immediately after he or she observed the tranquil scenes, and become aroused after being confronted by the erotic or disturbing. Naturally, study participants recorded the largest response once they'd seen the photos. However, what Radin discovered was that his subjects were also anticipating what they were about to see, registering physiological responses *before* they'd seen the photo. As if trying to brace themselves, their responses were highest before they saw an image that was disturbing. Blood pressure would drop in the extremities about a second before the image was flashed. Strangest of all, possibly reflecting that Americans are more unsettled about sex than violence, Radin discovered a far higher foreboding with the erotic than with the violent. He realized that he had some of the first laboratory proof that our bodies unconsciously anticipate and act out our own future emotional states. It also suggested that the 'nervous system is not just "reacting" to a future shock, but is also working out the emotional meaning of it'.[11]

Radin's studies were successfully replicated by his Dutch counterpart, a psychologist called Dick Bierman at the University of Amsterdam.[12] Bierman went on to use this model to determine whether people anticipate good or bad news. In studying the electrodermal activity of people involved in another published study which was examining learned response in a particular type of gambling card game, Bierman found that the participants registered rapid changes in EDA response *before* they were handed out their cards. Furthermore, these differences tended to correspond to the type of cards they got. Those about to receive a bad hand were more rattled and had all the hallmarks of a heightened fight-or-flight response.[13] This would seem to indicate that, on a subconscious physiological level, we have an inkling when we are about to receive bad news or when bad things are going to happen to us.

Radin tried another test of seeing into the future using a variation on Helmut Schmidt's machine. This type of machine was a 'pseudo random event generator', still unpredictable, but through a different mechanism. In this instance, a seed number, or initiating number, would kick off a highly complex mathematical sequence of other numbers. The machine contained 10,000 different seed numbers and so 10,000 different mathematical possibilities. The pseudorandom number generator was designed to produce sequences of random bits, or zeros and ones. Those sequences with the most 'ones' in them were deemed the best sequences and therefore the most desirable. The object was to stop the machine at a particular moment, on a particular seed number, to initiate the best sequences.

That, of course, was the trick of it. The window of selection was impossibly small; as the clock in the computer ticks 50 times a second, your correct seed number would flash up in 20 millisecond windows – ten times faster than the reaction times of human beings. To be successful at this, somehow you had to intuitively know that a good seed number was coming up and press the machine down precisely at that exact millisecond. As impossible as it sounded, this was exactly what Radin and his SRI boss, Ed May, did. Over hundreds of trials, Radin and May were somehow able to 'know' just when to hit the button to achieve the favorable sequence.[14]

Helmut Schmidt was consumed by a delicious possibility: the prospect of turning back time. He'd been thinking about how the effects he'd been seeing with machines seemed to defy space or causation. What began taking shape in Schmidt's mind was almost an absurdity of a question: whether a person attempting to affect the output of one of his machines could do so *after* it had been run. If a quantum state was as ethereal as a fluttering butterfly, did it matter when you tried to pin it down, so long as you were the first to attempt it – the first observer?

Schmidt rewired his REG to connect it to an audio device so that it would randomly set off a click, which would be taped to be heard in a set of headphones by either the left or right ear. He then turned on his machines and taperecorded their output, making sure that no one, including himself, was listening. A copy of the master tape was made, again with no one listening, and locked away. Schmidt also intermittently created tapes that were to act as controls, those where no one would ever try to affect its left–right clicks. As expected, when they were played, these control tapes had left and right ear clicks that were more or less evenly distributed.

Then, a day later, Schmidt got a volunteer to take one of the tapes home. His assignment was to listen to it and try to influence more of the clicks to come into his right ear. Later, Schmidt had his computer count up left and right clicks. His result seemed to defy common sense. What he found was that this influencer had changed the output of the machine, *just as if he'd been present when it was being recorded in the first place.* Furthermore, these results were just as good as his ordinary REG tests, as good as if someone had been sitting in front of the machine.

After carrying out a number of these tests, Schmidt realized that an effect was going on, but he didn't think his participants had changed the past, or erased a tape and made a new one. What seemed to have happened was that his influencers had changed what had happened in the first place. Their influence had reached back in time and affected the randomness of the machine *at the time it was first recorded.* They didn't change what *had* happened; they affected what would have happened in the first place. Present or future intentions act on initial probabilities and determine what events actually come into being.

Over more than 20,000 trials in five studies between 1971 and 1975, Schmidt showed that a highly significant number of tapes deviated from what was expected – roughly 50 per cent each of left and right clicks. He got similar results using machines that moved a needle on a dial, left or right. Of 832 runs, nearly 55 per cent had more left-hand needle moves than right.[15] Of all the studies on time travel, Schmidt's were probably the safest. Since a copy of the results had been made and locked away, it eliminated the possibility of fraud. What they showed decisively was that PK effects on a random system like a REG machine can occur at any time, past or future.

Schmidt also found that it was important for the influencer to be the first observer. If anyone else heard the tape first and listened with focused attention, the system seemed to make it less susceptible to influence later. Any form of focused attention seemed to freeze the system into final being. A few sparse studies even suggest that observation by any living system, human or even animal, seemed to successfully block future attempts at time-displaced influence. Although these types of studies have been thin on the ground, they accord with what we know about the observer effect in quantum theory. It suggests that observation by living observers brings things into some sort of set being.[16]

Bob Jahn and Brenda Dunne also began playing around with time in their own REG trials. In 87,000 of their experiments, they asked their volunteers to address their attention to the machine's operations anywhere from three days to two weeks *after* the machine had been allowed to run. Once they looked at the data, what they found was incredible. In every regard, this data was identical to the more conventional data they'd generated when their experimenters were attempting their influence at the time the machine was being run – the differences between women and men were still there and overall population distortions were the same. There was just one important difference. In the 'time-displaced' experiments, the volunteers were getting bigger effects than in the standard experiments every time they'd willed the machine to produce heads. However, because of the relatively small numbers, Jahn and Dunne had to deem this weird effect non-significant.[17]

A number of other investigators tried this kind of backward time travel to influence the gerbils running in activity wheels or the direction of people walking in the dark (and hitting a photobeam), or even cars hitting a photobeam in a tunnel in Vienna during the rush hour. The revolutions on the wheels and hits of the photobeam were converted into clicks, and taped, stored and played for the first time between one day and a week later to observers, who attempted to influence the gerbils to run faster or the people or cars to run into the beam more often. Another study attempted to see if a healer could retroactively influence the spread of blood parasites in rats. Braud had even done his own studies recording the EDA response of certain individuals and asking them to review their response and try to influence their own EDAs. Radin had carried out a similar study with EDA tapes and healers. Schmidt had studies where he'd tried to affect his own prerecorded breathing rate. All told, ten of the nineteen studies showed effects significantly different from chance – enough to indicate that something out of the ordinary was going on here.[18]

It was results like these that most troubled Hal Puthoff. The type of zero-point energy he was most familiar with was electromagnetic: a world of cause and effect, of order, of certain laws and limits – in this case, the speed of light. Things did not go backward or forward in time.

This body of experiments suggested three possible scenarios to him. The first was a vision of an utterly deterministic universe, where everything that was ever going to happen had already occurred. Within this universe of absolute fixed determinacy, people with premonitions were simply tapping into information, which was, on some level, already available.

The second possibility was perfectly explainable within known theoretic laws of the universe. Radin's opposite number, University of Amsterdam's Dick Bierman, believed you could account for precognition through a well-known quantum phenomenon known as retarded and advanced waves – the so-called Wheeler–Feynman absorber theory, which says that a wave can travel backward in time from the future to arrive at its source. What happens between two electrons is this. When one electron jiggles a bit, it sends out radiating waves into both the past and the future. The future wave, say, would hit a future particle, which would also wiggle, while sending out its own advanced and retarded waves. The two sets of waves from these two electrons cancel out, except in the region between them. The end result of a wave from the first traveling backward and the wave from the second traveling forward is an instantaneous connection.[19] In premonitions, Radin speculated, it could be that, on a quantum level, we are sending out waves to meet our own future.[20]

The third possibility, which perhaps makes the most sense, is that everything in the future already exists at some bottom-rung level in the realm of pure potential, and that in seeing into the future, or the past, we are helping to shape it and bring it into being, just as we do with a quantum entity in the present with the act of observation. An information transfer via subatomic waves doesn't exist in time or space, but is somehow spread out and ever-present. The past and present are blurred into one vast 'here and now' so your brain 'picks up' signals and images from the past or the future. Our future already exists in some nebulous state that we may begin to actualize in the present. This makes sense if we consider that all subatomic particles exist in a state of all potential unless observed – which would include being thought about.

Ervin Laszlo has proposed one interesting physical explanation for time-displacement. He suggests that the Zero Point Field of electromagnetic waves has its own substructure. The secondary fields caused by the motion of subatomic particles interacting with The Field are called 'scalar' waves, which are not electromagnetic and which don't have direction or spin. These waves can travel far faster than the speed of light – like

Puthoff's imagined tachyons. Laszlo proposes that it is scalar waves that encode the information of space and time into a timeless, spaceless quantum shorthand of interference patterns. In Laszlo's model, this bottom-rung level of the Zero Point Field – the mother of all fields – provides the ultimate holographic blueprint of the world for all time, past and future. It is this that we tap into when we see into the past or future.[21]

To take time out of the equation, as Robert Jahn suggests, we need to take separateness out of it. Pure energy as it exists at the quantum level does not have time or space, but exists as a vast continuum of fluctuating charge. We, in a sense, are time and space. When we bring energy to conscious awareness through the act of perception, we create separate objects that exist in space through a measured continuum. By creating time and space, we create our own separateness.

This suggests a model not unlike the implicate order of British physicist David Bohm, who theorized that everything in the world is enfolded in this 'implicate' state, until made explicit – a configuration, he imagined, of zero-point fluctuations.[22] Bohm's model viewed time as part of a larger reality, which could project many sequences or moments into consciousness, not necessarily in a linear order. He argued that as relativity theory says that space and time are relative and in effect a single entity (space-time) and if quantum theory stipulates that elements that are separated in space are connected and projections of a higher-dimensional reality, it follows that moments separated in time are also projections of this larger reality.

> Both in common experience and in physics, time has generally been considered to be a primary, independent and universally applicable order, perhaps the most fundamental one known to us. Now, we have been led to propose that it is secondary and that, like space, it is to be derived from a higher-dimensional ground, as a particular order. Indeed, one can further say that many such particular interrelated time orders can be derived for different sets of sequences of moments, corresponding to material systems that travel at different speeds. However, these are all dependent on a multidimensional reality that cannot be comprehended fully in terms of any time order, or set of such orders.[23]

If consciousness is operating at the quantum frequency level, it would also naturally reside outside space and time, which means that we theoretically have access to information, 'past' and 'future'. If humans are able to influence quantum events, this implies that we are also able to affect events or moments other than in the present.

This suggested one final intriguing thought to William Braud. Time-displaced human intention somehow acts on the probabilities of some occurrence to bring about an outcome, and works best on what Braud liked to call 'seed moments' – the first of a chain of events. So, if you applied these principles to physical or mental health, it could mean that we could use The Field to direct influences 'back in time' to alter pivotal moments or initial conditions which later bloom into full-blown problems or disease.

If thought in the brain is a probabilistic quantum process, as Karl Pribram and his colleagues propose, future intention might influence one neuron being fired and not another, setting off one or another chain of chemical and hormonal events that may or may not result in disease. Braud pictured a seed moment where a natural killer cell might exist in a 50–50 probabilistic state to kill or ignore certain cancer cells. That simple first decision might eventually make the difference between health and illness, or even death. There may be a score of ways that we could use intention in the future to change probabilities before they develop into full-blown disease. In fact, even the diagnosis itself might influence the future course of the disease and so should be approached with caution.

If the disease had developed, it wouldn't be that you could undo it. But some of the most harmful aspects of it might not have been actualized yet and might still be susceptible to change. You'd catch a disease at a point where it could be swayed in many directions, from good health to death. Braud pondered whether any cases of spontaneous remission had been caused by a future intention acting upon a disease before the point of no return. It might well be that every moment of our lives influences every other moment, forward and backward. As in *The Terminator* films, we might be able to go back in time to affect our own future.[24]

Part 3

Tapping into the Field

'The last century was the atomic age, but this one could well turn out to be the zero-point age.'

Hal Puthoff

CHAPTER TEN

The Healing Field

PUTHOFF, BRAUD AND THE other scientists had been left with an imponderable: the ultimate usefulness of the non-local effects they had observed. Their studies suggested a number of elegant metaphysical ideas about man and his relation to his world, but a number of practical considerations had been left unanswered.

How powerful was intention as a force and exactly how 'infectious' was the coherence of individual consciousness? Could we actually tap into The Field to control our own health or even to heal others? Could it cure really serious diseases like cancer? Was the coherence of human consciousness responsible for psychoneuroimmunology – the healing effect of the mind on the body?

Braud's studies in particular suggested that human intention could be used as an extraordinarily potent healing force. It appeared that we could order the random fluctuations in the Zero Point Field and use this to establish greater 'order' in another person. With this type of capability, one person should be able to act as a healing conduit, allowing The Field to realign another person's structure. Human consciousness could act as a reminder, as Fritz Popp believed, to re-establish another person's coherence. If non-local effects could be marshalled to heal someone, then a discipline like distant healing ought to work.

What was clearly needed was a test of these ideas in real life with a study so carefully designed that it would answer some of these questions, once and for all. In the early 1990s the opportunity arose with the perfect candidate – a scientist rather skeptical of the remote healing with a group of patients who'd been given up for dead.

Elisabeth Targ, an orthodox psychiatrist in her early 30s, was the daughter of Russell Targ, Hal Puthoff's partner and successor in the SRI remote-viewing experiments. Elisabeth was a curious hybrid, drawn to the possibilities suggested by her father's remote-viewing work at SRI, but also shackled by the rigors of her scientific training. At the time, she'd been invited to work as director of the California Pacific Medical Center's Complementary Research Institute, as a result of the remote-viewing work she'd done with her father. One of her tasks was to formally study the treatments offered by the clinic, which consisted largely of alternative medicine. Often she seemed to be teetering between both camps – wanting science to embrace and study the miraculous, and wanting alternative medicine to be more scientific.

A number of different strands in her life began to converge. She'd received a phone call from a friend of hers, Hella Hammid, announcing that she had breast cancer. Hella had arrived in Elisabeth's life through her father, who'd inadvertently discovered in Hella, a photographer, one of his most talented remote viewers. Hella had called to ask if Elisabeth knew of any evidence that alternative therapies such as distant healing – something not unlike remote viewing – could help to cure breast cancer.

In the 1980s, at the height of the AIDS epidemic – a time when a diagnosis of HIV was almost certainly a death sentence – Elisabeth had chosen this specialty in San Francisco, the very epicenter of the US epidemic. At the time of Hella's phone call, the hottest topic in medical circles in California was psychoneuroimmunology. Patients had begun to crowd into special town-hall meetings given by mind–body devotees such as Louise Hay or into workshops on visualization and imagery. Elisabeth herself had been dabbling in her own studies of mind–body medicine, undoubtedly because she'd had nothing much else to offer patients with advanced AIDS, even though she was deeply skeptical of Hay's approach. One of her own early studies had shown that group therapy was as good as Prozac for treating depression in AIDS patients.[1] She'd also read of the work of David Spiegel at Stanford Medical School, showing that group therapy dramatically increased life expectancy for women with breast cancer.[2]

In her sensible, pragmatic heart, Elisabeth suspected the effect was a combination of hope and wishful thinking, and perhaps a bit of confidence engendered by the support of the group. They may have been psychologically better, but their T-cell counts certainly weren't improving. Still, she harbored a shred of doubt, possibly derived from the years she'd spent observing her father's work on remote viewing at SRI. His success strongly argued for the existence of some sort of extrasensory connection between people and a field that connected all things. Elisabeth herself had often wondered if one could use the special ability observed in remote viewing for something besides spying on the Soviets or predicting a horse race, as she had once done.

Then in 1995, Elisabeth received a phone call from Fred Sicher. Fred was a psychologist, researcher and retired hospital administrator. He'd been referred to her friend Marilyn Schlitz, Braud's old associate, who was now the director of the Institute of Noetic Sciences, the Sausalito-based organization that Edgar Mitchell had set up many years before. Fred now at last had the time in his life to investigate something that fascinated him. As a hospital administrator, he'd always been something of a philanthropist. At Schlitz's suggestion, he approached Elisabeth about the possibility of working with him on a study of distant healing. With her unique background, Elisabeth was a natural choice to head up the study.

Prayer was not something Elisabeth had much experience of. She had inherited from her father not only her melancholic Russian looks and thick black tresses, lightly tinged with grey, but also her passion for the microscope. The only God in the Targ family home had been the scientific method. Targ had imparted to his daughter a sense of the thrill of science, with its capability of answering the big questions. As he'd chosen to work out how the world works, so his daughter had chosen to figure out the workings of the human mind. As a 13-year-old, she'd even wangled a place working in Karl Pribram's brain research laboratory at Stanford University, examining differences between left and right hemisphere activity, before deciding on an orthodox course of study in psychiatry at Stanford.

Nevertheless, Elisabeth had been highly impressed by the Soviet Academy of Science during a visit she'd made there with her father, and the fact that laboratory study of parapsychology could be so openly carried out by the establishment. In officially atheist Russia, they had only two categories of belief: something was true or not true. In America, a third category existed: religion, which placed some things strictly beyond the reach of scientific investigation. Everything scientists couldn't explain, everything connected with healing, or prayer, or the paranormal – the territory of her father's work – seemed to fall into this third category. Once it was placed there, it was officially declared out of bounds.

Her father had built his reputation on designing impeccable experiments, and he had taught her respect for the importance of the air-tight, well-controlled trial. She grew up believing that any sort of effect could be quantified, so long as you designed the experiment to control for variables. Indeed, Puthoff and Targ between them had demonstrated that the well-designed experiment could even prove the miraculous. The outcome was gospel, regardless of whether that outcome violated the researcher's every expectation. All good experiments 'work': the problem is simply that we may not like the conclusions.

Even as Targ senior shifted his thinking to embrace certain spiritual ideas, Elisabeth remained the cool rationalist. Still, throughout what was an orthodox training in psychiatry, she'd never forgotten her father's lessons: received wisdom was the enemy of good science. As a student she would seek out dusty psychiatric writings of the nineteenth century, before the advent of modern psychopharmacology, when psychiatrists lived in sanatoriums, writing down the rantings of their patients in an attempt to gain further understanding of their conditions. Somewhere in the raw data, Targ believed, separated out from the dogma of the times, lay the truth.

Elisabeth agreed to collaborate with Sicher, even though privately she doubted it was ever going to work. She would put distant healing to the purest test. She would try it out on her patients with advanced AIDS, a group so certain to die that nothing other than hope and prayer was open to them anymore. She would find out whether prayer and distant intention could cure the ultimate hopeless case.

She began trawling through the evidence on healing. The studies seemed to fall into three broad categories: attempts to affect isolated cells or enzymes; healing of animals, plants or microscopic living systems; and studies of human beings. Included was all of Braud and Schlitz's work, which showed that people could have an influence on all types of living processes. There was also some interesting evidence showing the effects humans could have on plants and animals. There'd even been some work showing that positive or negative thoughts and feelings could somehow be transmitted to other living things.

In the 1960s, biologist Dr Bernard Grad of McGill University in Montreal, one of the earliest pioneers in the field, was interested in determining whether psychic healers actually transmit energy to patients. Rather than using live human patients, Grad had used plants which he'd planned to make 'ill' by soaking their seeds in salty water, which retards growth. Before he soaked the seeds, however, he had a healer lay hands on one container of salt water, which was to be used for one batch of seeds. The other container of salt water, which had not been exposed to the healer, would hold the remainder of seeds. After the seeds were soaked in the two containers of salt water, the batch exposed to the water treated by the healer grew taller than the other batch.

Grad then hypothesized that the reverse might also happen – negative feelings might have a negative effect on the growth of plants. In a follow-up study, Grad had several psychiatric patients hold containers of ordinary water which were to be used again to sprout seeds. One patient, a man being treated for psychotic depression, was noticeably more depressed than the others. Later, when Grad tried to sprout seeds using water of the patients, *the water that had been held by the depressed man suppressed growth.*[3] This may be one good explanation why some people have green fingers and others can get nothing living to grow.[4]

In later experiments, Grad chemically analyzed the water by infrared spectroscopy and discovered that the water treated by the healer had minor shifts in its molecular structure and decreased hydrogen bonding between the molecules, similar to what happens when water is exposed to magnets. A number of other scientists confirmed Grad's findings.[5]

Grad moved on to mice, who'd been given skin wounds in the laboratory. After controlling for a number of factors, even the effect of warm hands, he found that the skin of his test mice healed far more quickly when healers had treated them.[6] Grad also showed that healers could reduce the growth of cancerous tumors in laboratory animals. Animals with tumors which were not healed died more quickly.[7] Other animal studies have shown that amyloidosis, tumors and laboratory-induced goiter could be healed in laboratory animals.[8]

Other conducted scientific studies had shown that people could influence yeast, fungi and even isolated cancer cells.[9] In one of them, a biologist named Carroll Nash at St Joseph's University in Philadelphia found that people could influence the growth rate of bacteria just by willing it so.[10]

An ingenious trial by Gerald Solfvin showed that our ability to 'hope for the best' might actually affect the healing of other beings. Solfvin created a series of complex and elaborate conditions for his test. He injected a group of mice with a type of malaria, which is usually fatal in rodents.

Solfvin got hold of three lab assistants and told them that only half the mice had been injected with malaria. A psychic healer would be attempting to heal one-half of the mice – not necessarily all those with malaria – although the assistants would not know which mice were to be the target of the healing. Neither statement was true.

All the assistants could do was to hope that the mice in their care would recover, and that the psychic healer's intervention would work.

However, one assistant was considerably more optimistic than his colleagues, and it showed. At the end of the study the mice under his care were less ill than those cared for by the other two assistants.[11]

Like that of Grad's healers, the Solfvin study was too small to be definitive. But there had been earlier research by Rex Stanford in 1974. Stanford had showed that people could influence events just by 'hoping' everything would go well, even when they did not fully understand exactly what they were supposed to be hoping for.[12]

Elisabeth was surprised to find that scores of studies – at least 150 human trials – had been done on healing. These were instances in which an intermediary would use one of a variety of methods to attempt to send healing messages, through touch, prayer or some sort of secular intention. With therapeutic touch, the patient is supposed to relax and attempt to direct his or her attention inward while the healer lays hands on the patient and intends the patient to heal.

A typical study involved ninety-six patients with high blood pressure and a number of healers. Neither doctor nor patients were told who was being given the mental healing treatments. A statistical analysis performed afterwards showed that the systolic blood pressure (that is, the pressure of blood flow as it is being pumped from the heart) of the group being treated by a healer was significantly improved, compared with that of the controls. The healers had employed a well-defined regime, which involved relaxation, getting in touch with a Higher Power or Infinite Being, using visualization or affirmation of the patients in a state of perfect health, and giving thanks to the source, whether it was God or some other spiritual power. As a group, the healers demonstrated overall success and, in certain individual instances, extraordinary results. Four of the healers enjoyed a 92.3 per cent improvement among their total group of patients.[13]

Perhaps the most impressive human study had been carried out by physician Randolph Byrd in 1988. It had attempted to determine in a randomized, double-blind trial whether remote prayer would have any effect on patients in a coronary care unit. Over 10 months, nearly 400 patients were divided into two groups, and only half (unbeknownst to them) were prayed for by a Christian prayer group outside the hospital. All patients had been evaluated, and there was no statistical difference in their condition before treatment. However, after treatment, those who'd been prayed for had significantly less severe symptoms and fewer instances of pneumonia and also required less assistance on a ventilator and fewer antibiotics than patients who hadn't been prayed for. [14]

Although a large number of studies had been carried out, the problem with many of them, as far as Elisabeth was concerned, was the potential for sloppy protocol. The researchers hadn't constructed trials tightly enough to demonstrate that it was truly healing that had produced the positive result. Any number of influences, rather than any actual healing mechanism, might have been responsible.

In the blood-pressure-healing study, for instance, the authors didn't record or control whether the patients were taking blood-pressure medication. Good as the results were, you couldn't really tell whether they were due to the healing or the drugs.

Although Byrd's prayer study was well designed, one obvious omission was any data concerning the psychological state of the patients when they'd started the study. As it is known that psychological issues can affect recovery after a number of illnesses, notably cardiac surgery, it may have been that a disproportionate number of patients with a positive mental outlook had landed in the healing group.

To demonstrate that healing was what had actually made patients better, it was vital to separate out any effects that might have been due to other causes. Even human expectation could skew the results. You needed to control for the effects of hope or such factors as relaxation on the outcome of trials. Cuddling animals, or even handling the contents of Petri dishes, could potentially bias the results, as could the act of traveling to a healer or even a warm pair of hands.

In any scientific trial, when you are trying to test the effectiveness of some form of intervention, you need to make sure that the only difference between your treatment group and control group is that one gets the treatment and the other doesn't. This means matching the two groups as closely as you can in terms of health, age, socioeconomic status and any other relative factors. If the patients are ill, you need to make sure that one group isn't more ill than the other. However, in the studies Elisabeth read about, few attempts had been made to make sure the populations were similar.

You also have to make sure that participation in a study and all the attention associated with it doesn't itself cause improvement, so that you have the same results among those who have been treated and those who haven't.

In one such study, a six-week distant healing study of patients suffering from clinical depression, the test was unsuccessful – all the patients improved, even the control group who hadn't been subject to healing. But all patients, those receiving healing and those with no healing, may have had a psychological boost from the session, which might have overwhelmed any actual effect of healing.[15]

All these considerations represented a tremendous challenge to Elisabeth in putting together a trial. The study would have to be so tightly constructed that none of these variables affected the results. Even the presence of a healer at certain times and not others might tend to influence the outcome. Though a laying on of hands might aid in the healing process, to control properly in a scientific sense meant that patients should not know whether they were being touched or healed.

Targ and Sicher spent months designing their trial. Of course, it had to be double-blind, so that neither patients nor doctors could know who was being healed. The patient population had to be homogeneous, so they selected advanced AIDS patients of Elisabeth's with the same degree of illness – the same T-cell counts, the same number of AIDS-defining illnesses. It was important to eliminate any element of the healing mechanism that might confound the results, such as meeting the healer or being touched. This meant, they decided, that all healing should be done remotely. Because they were testing healing itself, and not the power of a particular form of it, such as Christian prayer, their healers should be from diverse backgrounds and between them cover the whole array of approaches. They would screen out anyone who appeared overly egotistical, only in it for the money or fraudulent. They'd also have to be dedicated, as they'd receive no pay and no individual glory. Each patient was to be treated by at least ten different healers.

After four months of searching, Fred and Elisabeth had their healers – an eclectic assortment of forty religious and spiritual healers all across America, many highly respected in their fields. Only a small minority described themselves as conventionally religious and carried out their work by praying to God or using a rosary: several Christian healers, a handful of evangelicals, one Jewish kabbalist healer and a few Buddhists. A number of others were trained in non-religious healing schools, such as the Barbara Brennan School of Healing Light, or worked with complex energy fields, attempting to change colors or vibrations in a patient's aura. Some used contemplative healing or visualizations; others worked with tones and planned to sing or ring bells on behalf of the patient, the purpose of which, they claimed, was to reattune their chakras, or energy centers. A few worked with crystals. One healer, who'd been trained as a Lakota Sioux shaman, intended to use the Native American pipe ceremony. Drumming and chanting would enable him to go into a trance during which he would contact spirits on the patient's behalf. They also enlisted a Qigong master from China, who said that he would be sending harmonizing *qi* energy to the patients. The only criterion, Targ and Sicher maintained, was that the healers believed that what they were using was going to work.

They had one other common element: success in treating hopeless cases. Collectively, the healers had an average of 17 years of experience in healing and reported an average of 117 distant healings apiece.

Targ and Sicher then divided their group of twenty patients in half. The plan was that both groups would receive the usual orthodox treatment, but only one of the two groups would also receive distant healing. Neither doctors nor patients were going to know who was being healed and who wasn't.

All information about each patient was to be kept in sealed envelopes and handled individually through each step of the study. One of the researchers would gather up each patient's name, photograph and health details into a numbered folder. This would then be given to another researcher, who would then renumber the folders at random. A third researcher would then randomly divide the folders into two groups, after which they were placed in locked filing cabinets. Copies in five sealed packets would be sent to each healer, with information about the five patients and a start date specifying the days to begin treatment on each person. The only participants in the study who were going to know who was being healed were the healers themselves. The healers would have no contact with their patients – indeed, would never even meet. All they'd been given to work with was a photo, a name and a T-cell count.

Each of the healers was asked to hold an intention for the health and well-being of the patient for an hour a day, six days each week, for ten weeks, with alternate weeks off for rest. It was an unprecedented treatment protocol, where every patient in the treatment group would be treated by every healer in turn. To remove any individual biases, healers had a weekly rotation, so that they were assigned a new patient each week. This would enable all of the healers to be distributed throughout the patient population, so that healing itself, not any particular variety of it, would be studied. The healers were to keep a log of their healing sessions with information about their healing methods and their impressions of their patients' health. By the end of the study, each of the treated patients would have had ten healers, and each of the healers, five patients.

Elisabeth was open-minded about it, but the conservative in her kept surfacing. Try as she might, her training and her own predilections kept surfacing. She remained fairly convinced that Native American pipe smoking and chakra chanting had nothing to do with curing a group of men with an illness so serious and so advanced that they were virtually certain to die.

And then she saw her patients with end-stage AIDS getting better. During the six months of the trial period, 40 per cent of the control population died. But all ten of the patients in the healing group were not only still alive but had become healthier, on the basis of their own reports and medical evaluations.

At the end of the study, the patients had been examined by a team of scientists, and their condition had yielded one inescapable conclusion: the treatment was working.

Targ almost didn't believe her own results. She and Sicher had to make certain that it was healing that had been responsible. They checked and rechecked their protocol. Was there anything about the treatment group that had been different? Had the medication been different, the doctor different, their diets different? Their T-cell counts had been the same, they had not been HIV positive for longer. After re-examining the data, Elisabeth discovered one difference they'd overlooked: the control patients had been slightly older, a median age of 45, compared with 35 in the treatment group. It didn't represent a vast difference – just a ten-year age gap – but that could have been factor in why more of them had died. Elisabeth followed up the patients after the study, and found that those who'd been healed were surviving better, regardless of age. Nevertheless, she knew they were dealing with a controversial field and an effect that is, on its face, extremely unlikely, so science dictates that you have to assume the effect isn't real unless you are really sure. Occum's razor. Select the simplest hypothesis when confronted with several possibilities.

Elisabeth and Sicher decided to repeat the experiment, but this time to make it larger and to control for age and any other factors they'd overlooked. The forty patients chosen to participate were now perfectly matched for age, degree of illness and many other variables, even down to their personal habits. The amount they smoked, or exercise they took, their religious beliefs, even their use of recreational drugs were now equivalent. In scientific terms, this was a batch of men who were as close as you could get to a perfect match.

By this time protease inhibitors, the great white hope drug of AIDS treatment, had been discovered. All of the patients were told to take standard triple therapy for AIDS (protease inhibitors plus two anti-retroviral drugs such as AZT) but to continue their medical treatment in every other regard.

Because the triple therapy appeared to be making a profound difference on mortality rates in AIDS patients, Elisabeth assumed that, this time, no one in either group would actually die. This meant she needed to change the result she was aiming for. In the new study, she was looking for whether distant healing could slow down the progression of AIDS. Could it result in fewer AIDS-defining illnesses, improved T-cell levels, less medical intervention, improved psychological well-being?

Elisabeth's caution finally paid off. After six months, the treated group were healthier on every parameter – significantly fewer doctor visits, fewer hospitalizations, fewer days in hospital, fewer new AIDS-defining illnesses and significantly lower severity of disease. Only two of those in the treatment group had developed any new AIDS-defining illnesses, while twelve of the control group had, and only three of the treated group had been hospitalized, compared with twelve of those in the control group. The treated group also registered significant improved mood on psychological tests. On six of the eleven medical outcome measures, the group treated with healing had significantly better outcomes.

Even the power of positive thinking among the patients had been controlled for. Midway through the study, all the participants were asked if they thought they were being treated. In both the treatment and the control groups, half thought they were, half thought not. This random division of positive and negative views about healing meant that any involvement of positive mental attitude would not have affected the results. When analyzed, the beliefs of the participants about whether they were getting healing treatment did not correlate with anything. Only at the end of the study period did the subjects tend to guess correctly that they'd been in the healing group.

Just to be sure, Elisabeth conducted fifty statistical tests to eliminate whether any other variables in the patients might have contributed to the results. This time, there were no more than chance.

The results were inescapable. No matter which type of healing they used, no matter what their view of a higher being, the healers were dramatically contributing to the physical and psychological well-being of their patients.[16]

Targ and Sicher's results were vindicated a year later, when a study entitled MAHI (Mid-America Heart Institute) of the effect of remote intercessory prayer for hospitalized cardiac patients over 12 months showed patients had fewer adverse events and a shorter hospital stay if they were prayed for. In this study, however, the 'intercessors' were not gifted healers; to qualify to take part, they simply needed to believe in God and the fact that He responds when you pray to Him to heal someone who is ill. In this instance, all the participants were using some form of standard prayer and most were Christian–Protestant, Roman Catholic or nondenominational. Each was given a particular patient to pray for.

After a month, symptoms in the prayer group had been reduced by more than 10 per cent compared with those receiving standard care, according to a special scoring system developed by three experienced cardiologists from the Mid-America Heart Institute, which rates a patient's progress from excellent to catastrophic. Although the healing didn't shorten their hospital stay, the patients being prayed for were definitely better off in every other regard.[17]

More studies are now under way in several universities. Elisabeth herself began a trial (which, at the time of writing in 2001, is still going on) comparing the effects of distant healers with nurses, a group of health professionals whose caring attitude toward their patients might also act as a healing mechanism.[18]

The MAHI study offered several important improvements over Randolf Byrd's study. Whereas all the medical staff in Byrd's study had been aware that a study was being carried out, the medical staff in the MAHI study had no idea.

The MAHI patients also didn't know they were participating in a study, so there would not have been any possible psychological effects. In Byrd's study, of the 450 patients, nearly an eighth had refused to be involved. This meant that only those who were receptive to, or at least didn't object to the idea of, being prayed for would have agreed to be included. Finally, in Byrd's study, those doing the praying had been given a great deal of information about their patients, whereas in the MAHI study, the

Christians had virtually no information about the people they were praying for. They were told to pray for 28 days, and that was it. They had no feedback about whether their prayers had worked.

Neither the Targ nor the MAHI study demonstrated that God Himself answers prayers or even that He exists. As the MAHI study was quick to point out: 'All we have observed is that when individuals outside of the hospital speak (or think) the first names of hospitalized patients with an attitude of prayer, the latter appeared to have a "better" CCU experience.'[19]

In fact, in Elisabeth's study, it didn't seem to matter what method you used, so long as you held an intention for a patient to heal. Calling on Spider Woman, a healing grandmother star figure common in the Native American culture, was every bit as successful as calling on Jesus. Elisabeth began to analyze which healers had the most success. Their techniques had been profoundly different. One 'flow alignment' practitioner based in Pittsburgh felt, after attempting work with several of the patients, that there was a common energy field in all of them, which she came to think of as an 'AIDS energy signature', and she would work on getting in touch with their healthy immune system and ignore the 'bad energy'. With another it was more a case of working on psychic surgery, spiritually removing the virus from their bodies. Another, a Christian in Santa Fe, who carried out the healing in front of her own altar with pictures of the Virgin and saints and many lit candles, claimed to have summoned up spirit doctors, angels and guides. Others, like the kabbalistic healer, simply focused on energy patterns.[20]

But what they all seemed to have in common was an ability to get out of the way. It seemed to Elisabeth that most of them claimed to have put out their intention and then stepped back and surrendered to some other kind of healing force, as though they were opening a door and allowing something greater in. Many of the more effective ones had asked for help – from the spirit world or from the collective consciousness, or even from a religious figure such as Jesus. It was not an egoistic healing on their part, more like a request: 'please may this person be healed'. Much of their imagery had to do with relaxing, releasing or allowing the spirit, light or love in. The actual being, whether it was Jesus or Spider Woman, appeared irrelevant.

The success of the MAHI study suggested that healing through intention is available to ordinary people, although the healers may be more experienced or naturally talented in tapping into The Field. In the Copper Wall Project in Topeka, Kansas, a researcher named Elmer Green has shown that experienced healers have abnormally high electric field patterns during healing sessions. In his test, Green enclosed his participants in isolated rooms made with walls constructed entirely of copper, which would block electricity from any other sources. Although ordinary participants had expected electrical readings related to breathing or heartbeat, the healers were generating electrical surges higher than 60 volts during healing sessions, as measured by electrometers placed on the healers themselves and on all four walls. Video recordings of the healers showed these voltage surges had nothing to do with physical movement.[21] Studies of the nature of the healing energy of Chinese Qigong masters have provided evidence of the presence of photon emission and electromagnetic fields during healing sessions.[22] These sudden surges of energy may be physical evidence of a healer's greater coherence – his ability to marshall his own quantum energy and transfer it to the less organized recipient.

Elisabeth's study and the work of William Braud raised a number of profound implications on the nature of illness and healing. It suggested that intention on its own heals, but that healing is also a collective force. The manner in which Targ's healers worked would suggest that there may be a collective memory of healing spirit, which could be gathered as a medicinal force. In this model, illness can be healed through a type of collective memory. Information in The Field helps to keep the living healthy. It might even be that health and illness of individuals is, in a sense, collective. Certain epidemics might grip societies as a physical manifestation of a type of energetic hysteria.

If intention creates health – that is, improved order – in another person, it would suggest that illness is a disturbance in the quantum fluctuations of an individual. Healing, as Popp's work suggests, might be a matter of reprogramming individual quantum fluctuations to operate more coherently. Healing may also be seen as providing information to return the system to stability. Any one of a number of biological processes requires an exquisite cascade of processes, which would be sensitive to the tiny effects observed in the PEAR research.[23]

It could also be that illness is isolation: a lack of connection with the collective health of The Field and the community. Indeed, in Elisabeth's study, Deb Schnitta, the flow alignment practitioner from Pittsburgh, found that the AIDS virus seemed to feed on fear – the type of fear that might be experienced by anyone shunned by the community, as many homosexuals were during the beginning of the AIDS epidemic. Several studies of heart patients have shown that isolation – from oneself, one's community and one's spirituality – rather than physical conditions, such as a high cholesterol count, is one of the greatest contributors to disease.[24] In studies of longevity, those people who live longest are often not only those who believe in a higher spiritual being, but also those who have the strongest sense of belonging to a community.[25]

It might mean that the intention of the healer was as important as his or her medicine. The frantic doctor who wishes his patient could cancel so he could have his lunch; the junior doctor who has stayed up for three nights straight; the doctor who doesn't like a particular patient – all may have a deleterious effect. It might also mean that the most important treatment any doctor can give is to hope for the health and well being of his or her patient.

Elisabeth began to examine what was present in her consciousness just before she went in to see her patients, to make sure that she was sending out positive intentions. She also began to study healing. If it could work for Christians who didn't know the patients they were praying for, she thought, it could also work for her.

The *modus operandi* of her healers suggested the most outlandish idea of all: that individual consciousness doesn't die. Indeed, one of the first serious laboratory studies of a group of mediums by the University of Arizona seems to validate the idea that consciousness may live on after we die. In studies carefully controlled to eliminate cheating or fraud, the mediums typically were able to produce more than eighty pieces of information about deceased relatives, from names and personal oddities to the actual and detailed nature of their deaths. Overall, the mediums achieved an accuracy rate of 83 per cent – and one had even been right 93 per cent of the time. A control group of non-mediums were only right, on average, 36 per cent of the time. In one case, a medium was able to recite the prayer a deceased mother used to recite for one of the sitters as a child. As Professor Gary Schwarz, who led the team, said, 'The most parsimonious explanation is that the mediums are in direct communication with the deceased.'[26]

As Fritz-Albert Popp described it, when we die we experience a 'decoupling' of our frequency from the matter of our cells. Death may be merely a matter of going home or, more precisely, staying behind – returning to The Field.

CHAPTER ELEVEN

Telegram from Gaia

IT HAD TO BE the most gripping moment Dean Radin could think of, and nothing, he decided, was more gripping than the end of the O.J. Simpson trial, which had overtaken the Stopes 'monkey' trial as the American trial of the century. From the moment that the white Ford Bronco had skittishly raced along the LA freeway, tens of millions of Americans per minute had watched the drama unfold on court TV. And now, nearly a year into the trial, half a billion viewers worldwide had turned on their television sets, ready to watch the live broadcast of the fate of the Bronco's driver, who was awaiting the jury's verdict as to whether he had or had not brutally slashed to death his wife and her lover.

So many Americans had remained riveted to their television sets throughout the nine and a half months of the trial, the 133 days of testimony, the 126 witnesses, the 857 exhibits entered into evidence, the issues of racism, the DNA testing and bloody gloves, the staggering blunders of the police and forensic experts, the drama when Judge Lance Ito twice threw out the television cameras and roundly chastised the two squabbling legal teams, that it had cost the American gross national product an estimated $40 billion in lost productivity. And now a year and four days after the jury had first been selected, this true-life drama which had made for so much compulsive viewing, which had cut so deeply into daytime soap opera viewing that it could command its own premium television advertising space, was about to come to an end.

Even the final moments had their unexpected dramatic cliffhanger. Just as the jury had reached their verdict and were assembled in the courtroom, Armanda Cooley, the jury foreman, realized that she'd left the form with their verdict written on it, sealed in its envelope, in the jury room. But even if she'd had it there, two lawyers for the defense, including Johnny Cochran, the head of Simpson's 'dream team' of prominent attorneys, weren't present. Judge Ito declared a recess. The verdict would be read the following morning at 10 a.m. The world would have to wait one more day.

On October 3, 1995, an audience greater than that for three of the five previous Superbowls or for the 'Who shot JR?' episode of *Dallas* turned on its television sets. Judge Ito asked that the verdict be passed to the court clerk, Deirdre Robertson. She and O.J. Simpson stood up. The world held its breath.

'In the matter of People of the State of California vs Orenthal James Simpson, case number BA 097211. We, the jury, in the above-entitled action, find the Defendant, Orenthal James Simpson, not guilty,' read Mrs Robertson.

O.J. Simpson, so impassive through most of the trial, broke into a triumphant smile.

O.J. was cleared on both counts. It was the final twist in the tale. The television audience was stunned by the jury's decision, and so were five other silent observers – all REG computers, one at the PEAR lab, another at the University of Amsterdam and three more at the University of Nevada. They'd been set to run continuously for three hours before, during and after the reading of the verdict.

Afterwards, Radin examined their output. Three statistically significant peaks of highs had occurred in all five computers at exactly the same three moments: a small peak at 9 a.m. Pacific time, a larger peak an hour later, and then an enormous peak seven minutes after that. These three blips corresponded to the three most important final moments of the trial: when the show first started, with the initial television commentary – the time when most people would have turned on their television sets – then the beginning of the broadcast of the actual courtroom proceedings, and finally the exact moment the verdict was announced. Like everyone else in the world, these computers had snapped to attention to find out whether O.J. was innocent or guilty.[1]

The possibility that a collective consciousness might exist had been taking shape for many years in Dean Radin's mind, perhaps even influenced by his mother, who'd been interested in yoga all those years ago. Certainly, this notion was a familiar concept in ancient and Eastern cultures. But others, like psychologist William James, had proposed that the brain simply reflects this collective intelligence, like a radio station picking up signals and transmitting them. As Radin and his colleagues observed the apparent ability of the human mind to extend its boundaries, natural questions arose about whether the effects get larger when many individuals operate in unison and indeed whether a collective global mind ever operated as a unity. If coherence could develop between individuals and their environment, was there also a possibility of group coherence?

What was different about Radin's thoughts was that he was trying to work out how to test it scientifically. It was Roger Nelson who had first thought to see if a REG machine could pick up evidence of a collective consciousness. The idea grew out of an experience he'd had one day while he was studying some data at the PEAR lab. It was 1993 and Nelson was a 53-year-old doctor of psychology, unofficially looked upon as the coordinator of experiments at the PEAR lab, a natural hand at directing, the fellow who got everybody together to make sure the job got done. He'd come to the lab in 1980 for a year-long sabbatical from teaching at a college in Vermont, but then one year turned into two, and before long he informed his college that he wasn't coming back. The PEAR work was intoxicating for the Nebraska-born Nelson, red-bearded and rustic-featured, another philosopher scientist drawn, even as a child, to the scientific frontier.

Nelson had been sitting up in the civil engineering department at Princeton, creating graphs for the distributions of the scores for multiple REG runs. As he examined the graphs for runs where people had put out one set of intentions (HIs) and graphs for the opposite intention (LOs), there was nothing out of the ordinary. As expected, the graph of the HIs was shifted a little to the left, and that of the LOs was shifted a little to the right. Roger then pulled up the statistics for the third test, when people were not supposed to have any intention toward the machine. It was supposed to be a baseline, with a shape that was virtually indistinguishable from those of pure chance when the machine was running by itself, with nobody trying to affect it. The graph was nothing like that. It was all squeezed together. In the very center, there was a neat and obvious exception, a little bar jutting up, resembling nothing so much as a clenched little fist. There it was, wagging at him in reproach. Nelson laughed so hard at it that he fell off his chair. How could he have failed to recognize this? Even trying not to think of anything might create its own focus of energy. Your mind couldn't help it. Intending not to have any effect on a REG machine was like trying not to think of elephants. Perhaps any sort of attention, by its very act of focusing consciousness, could create order. The mind was always carrying on – noticing, thinking.

We think, therefore we affect.

There had already been some evidence of this in the PEAR lab. Nelson had seen that certain people, often women, had more dramatic success in influencing the REG machines when they were concentrating on something else.[2] Nelson began by testing this with a device he'd named ContREG – shorthand for keeping a REG machine running continuously to see if it registered any more heads or tails than usual in the ordinary course of the day and then establishing what had been going on in the room during the moments of effect.

Out of that grew another idea. Everyday observing requires a very low state of attention. You take in many sights, sounds and smells around you in the course of your ordinary activities. However, when you do something that really engages your mind and emotions – listening to music, watching a gripping moment of theater, attending a political rally or a religious service – you concentrate with every pore of your body. You attend to it in a state of peak intensity.

Nelson wondered first whether the ability of consciousness to order or influence depends upon how intent the observer is. And second, if it does for individuals, what would be the effect of more than one person? He'd seen from the PEAR data that bonded couples – people who were intensely involved – had a more profound effect on the REG machines than individuals. It suggested that two like-minded people created more order in a random system. Suppose you assemble an entire crowd, all focusing intently on the same thing. Would the effect be even greater? Was there a relation between the size of the crowd or the intensity of interest and the size of the effect? After all, he thought, everyone had had moments in their lives where the consciousness of a group event could almost be felt. A REG machine was so exquisitely sensitive that it might just pick up on this.

Nelson decided to test out this theory with meetings that were to hand. Robert Jahn and Brenda Dunne were already planning to attend the International Consciousness Research Laboratories in April 1993, where a group of senior scholars met twice a year to exchange information about the role of consciousness. Later that year, Nelson planned to attend the Direct Mental Healing Interactions (DHML) group, held at the Esalen Institute in California, which promised to be a powerful conference of a dozen scientists examining how to conduct research on healing. In Hollywood, a certain awe was reserved for people who were 'good meetings'. In Nelson's case, the question was whether a REG machine would pick up the good vibrations as well.

Jahn and Dunne headed off to their meeting with a box and a laptop computer, which represented the REG program and the computer recording the data, and kept it running throughout their conference. Nelson did the same at his Esalen meeting. What they were looking for was whether this steady shift from random movement would indicate some change in the 'information' environment and be related to the shared information field and collective consciousness of the group.[3] The main difference between these and the ordinary REG trials was that the group wouldn't be trying to influence the machine in any way.

When they all returned to Princeton and analyzed the results, they discovered that some undeniable effect had taken place. They decided to carry out a series of these experiments. At another, similar event – this time, the Academy of Consciousness sponsored by ICRL – the data was even more decisive. A big central incline in the graph corresponded exactly with the point during the meeting where there'd been an intense, twenty-minute discussion concerning ritual in everyday life, which had captivated the audience. Nelson also examined log books and audio recordings of group members made at the time. Many of the fifty attendees had remarked upon the discussion as a special shared moment. Without knowing of the outcome of the REG machine, one member had reported that a change in the group's energy had been almost palpable.[4]

With his own Esalen study, Nelson discovered that the most riveting moment of the meeting had also produced a strong deviation from randomness in the data.

The results were intriguing, but the idea needed to be tested further, in all sorts of venues. To best accomplish this, though, he needed a device that was truly portable. The hardware had been cumbersome and unwieldy, requiring its own power supply. Nelson thought of using a Hewlett Packard palm computer, which was not much bigger than a pocket tape recorder, with a miniaturized REG device sitting on top, plugged into the serial port, kept in place with a piece of Velcro.

Nelson wasn't interested in whether he'd got more heads than tails since no one would be expressing an intention. All he wanted to determine was whether the machine had deviated in any direction away from its 50–50 random activity. Any change – whether more heads or more tails, or sometimes more heads and then sometimes more tails – would be construed as a departure from chance. This called for a different statistical method of analyzing the data from that used by the PEAR lab for its ordinary studies. Nelson decided to use a method called 'chi square', which entailed plotting the square of each individual run. Any unusual behavior, some prolonged or extreme deviation from its expected random heads-or-tails-type monotony, would easily show up.

Nelson had called these experiments in 'field consciousness', or 'Field-REG', for short. The name had had a neat double entendre. It was a REG out in the field, but also a device used to test if there was such a thing as a 'consciousness field'.

Nelson decided to try his FieldREG on events of every variety – business meetings, academic meetings, a humor conference, concerts, theatrical events. He sought out compelling events that would keep the audience riveted – moments when a great number of people were all engaged in the same intense thought at the same time.[5] When a member of the Covenant of Unitarian Universalist Pagans (CUUPS) expressed an interest in the PEAR work, Nelson loaned him a FieldREG and the machine attended fifteen of their ritual pagan gatherings – including Sabbats and those held during the full moon.[6]

The friend of a PEAR colleague, the artistic director of a large musical review called *The Revels*, which is mounted in eight US cities each December to see in the New Year, approached Nelson about trying out a FieldREG trial with his show. It seemed perfect: it had ritual, it had music, it had audience participation. Roger viewed the production and asked the artistic director to pick the five most engaging portions of the show that would most affect the audience and hence the machine. The FieldREG attended ten shows in two cities in 1995 and several performances in eight cities in 1996. As if on cue, each moment that Nelson had predicted caused a glitch in the machine's data.[7]

A definite pattern was emerging. The machine was moving out of its random movements into some sort of order precisely during moments of peak attention: special presentations at meetings, the climaxes of humor conferences, the most intense moments of a pagan ritual. For a REG machine, whose movements were so delicately minuscule, these effects were relatively large – three times what it was for individuals at PEAR trying to affect the machines on their own. In the pagan sessions, the FieldREG had veered wildly off course twice, both during full-moon rituals, recording many more tails than usual.

One CUUPS group member was not surprised when Nelson told him the results. 'On the whole,' he remarked, 'our Sabbats are not very personal or intense, whereas the moons sometimes are.'[8]

The particular activity didn't really matter. What seemed most important was the intensity of the group, the ability of the activity to keep its audience spellbound, and it helped if there was some sort of collective resonance in the group, particularly some context that was emotionally meaningful to them. At the humor conference, the machine made its biggest deviation during an evening keynote presentation, which was so funny the audience had given the comic a standing ovation and demanded an encore. What was clearly most important was that everyone was focused in rapt attention, all thinking the same thought.

What appeared to be happening was that when attention focused the waves of individual minds on something similar, a type of group quantum 'superradiance' occurred which had a physical effect. The REG machine was in a sense a kind of thermometer, measuring the dynamics and coherence of the group. Only the business and academic meetings had no effect on the machine. If a group was bored and its attention was wandering, in a manner of speaking the machine was bored, too. It was just the intense moments of like-mindedness which seemed to gather enough power to impart some order on the chaotic purposelessness of a REG machine.

The idea of sacred sites intrigued Nelson. Were they sacred because their use over the centuries had invested them with that quality, or had there been a quality about the site – the configuration of trees or stones, the spirit of place, its very location – that had been there from the beginning, leading human beings to naturally select it for that purpose? Ancient peoples had been sensitive to the earth's signals, able to read and pay attention to certain configurations such as ley lines. If there was something different about the place itself, had a type of collective consciousness coalesced there like an energetic whorl, or had some sort of energetic resonance always existed? And would any of this register on a REG machine?

Nelson decided to seek out several sites in America that had been sacred to Native Americans. Nelson and his machine observed a medicine man performing a ritual healing ceremony at the Devil's Tower monument in Wyoming, a place considered sacred by certain tribes. Later, he walked around Devil's Tower himself with a PalmREG in his pocket, and then visited Wounded Knee in South Dakota, the site of the massacre of an entire Sioux tribe. Nelson surveyed the desolation, the cemetery and the monument to the dead. He fell into a deep quiet. Later, when he looked at the data for the two places, it was beyond doubt: his machine's output was definitely being affected, and with a far larger effect size than ordinary PEAR studies, as though there were some lingering memory of the thoughts of all the people who'd lived and died there.[9]

The perfect opportunity to look closer at the nature of collective memory and resonance arose during a trip to Egypt. Nelson decided to attend a two-week tour of Egypt with a group of nineteen colleagues, planning to visit the main temples and sacred sites of the ancient Egyptians, where they would carry out a series of informal ceremonies, such as chanting and meditation. This trip would give him the chance to see whether people engaged in meditative activities at these sites – the kind of activities, in a sense, for which the sites had originally been built – had even more effect on the machines. Nelson kept a PalmREG running in his coat pocket during visits to all the major sites – the great Sphinx, the Temples of Karnak and Luxor, the Great Pyramid of Giza. The PalmREG was on while the group meditated or chanted and when they were simply wandering through the temples, and even during moments when he was on his own, touring or meditating. He also kept a careful record of times when various activities had occurred.

When he'd returned home and compiled all his data, an interesting pattern emerged. The strongest effects on the machine occurred during times when the group was engaged in a ritual such as chanting at a sacred site. In most of the main pyramids, the effects had been six times that of ordinary REG trials at PEAR and twice those of ordinary FieldREG trials. These were among the largest effects he'd seen – as large as those for a bonded couple. But when he put together all the data of the twenty-seven sacred sites he'd visited, while simply walking around them with no more than a respectful silence, the results were even more astounding. The spirit of the place itself appeared to register effects every bit as large as the meditating group.

Of course, as he was carrying around the PalmREG in his pocket, his own expectations might have affected it – a well-known phenomenon referred to as the 'experimenter effect'. It could have been the collective expectations and awe of the other visitors – after all, he was never at the sites on his own. But some other controls demonstrated that the situation was a little more complicated. Again, when the group attempted chanting and meditation in other sites which were not deemed to be sacred but were nevertheless interesting, the effects on the PalmREG were significant, but smaller. Even when the members of the group seemed attuned to each other – during a solar eclipse, attending a special astrology session, or a sunset birthday party – the machine's effects were also small, not much greater than the effects observed during a standard REG trial. Nelson even monitored a series of his own focused ritual – during prayer at a mosque or certain ritual walks and while observing and trying to 'decode' hieroglyphics. Many of them had been involving to Nelson – some deeply moving. Nevertheless, the machine's output deviated a little, but no more than it would have if he were home in Princeton, sitting in front of a REG machine. Clearly, some resonance reverberated at the sites, possibly even a vortex of coherent memory.

Both the type of place and the activity of the group seemed to play contributing roles in creating a kind of group consciousness. In the sacred sites where chanting hadn't taken place, simple group presence, or perhaps even the place itself, held a high degree of resonating consciousness. The machine had also registered an effect, even in the midst of the more mundane activities or places, so long as the group's attention had been aroused. And no matter how deeply engaged Nelson had been on his own, he could not match the effect size of the group.

There was one other remarkable element of his data. During his trip to the Great Pyramid of Khufu on the Giza plateau, the PalmREG had veered off its random course with a positive trend during two group chants inside the Queen's Chamber and the Grand Gallery and then had a strongly negative trend in the King's Chamber, where they'd carried on their chant. A similar situation had occurred at Karnak. Nelson was amazed once the results had been plotted on a graph; both of them formed a large pyramid. It was hard to keep from thinking that, on some level, the PalmREG had been experiencing Nelson's trip in parallel.[10]

Dean Radin had been at the Direct Mental Healing meeting and had seen Nelson's weird data. As Radin had been an associate of Nelson's and a co-author of the PEAR data meta-analysis, he was a natural candidate to replicate Nelson's work.

With his first studies, Radin, like Nelson, discovered that these effects happen when a FieldREG is present in the room or at the site. But what about at long distance? The most obvious vehicle for long-distance like-mindedness was television. Everybody watched television, particularly the popular shows. Would they all be thinking the same thing while they watched? To test this, Radin needed something beyond a sitcom – an event that would guarantee an audience on the edge of its seat.[11] The O.J. Simpson trial verdict would later represent a natural choice. But for his first study, Radin chose the Sixty-seventh Academy Awards in March 1995, which, with its estimated viewer size of one billion, was one of the biggest audiences he could think of. This audience comprised people in 120 different countries, so their contribution in mass attention would be coming from around the world.

To further demonstrate that the effects happened instantaneously at any distance, Radin used two REG machines, placed in different spots. One sat about 20 yards from him as he watched the event on March 27, the other was in his lab about 12 miles away, running on its own and not in front of a television set. During the broadcast, both Radin and his assistant painstakingly noted down, minute by minute, the high interest and low interest moments of the show. Any moments of peak tension, such as the announcement of the winners for best picture, best actor or actress, were timed and noted as 'high coherence' periods.

After the show ended, he examined his data. During the highest interest periods, the machines' degree of order increased to such a level that the odds against it having occurred by chance were 1000 to 1. During the low interest periods, on the other hand, the degree of order was at a lower level, with odds against it having occurred by chance no greater than 10 to 1. Both computers were also run for four hours after the event, and during this control period, after a tiny high, possibly reflecting the end of the awards ceremony, both quickly returned to their usual random behavior. Radin replicated his own experiment a year later, with similar results. He got the same kind of results with the Summer Olympics of July 1996 and of course the O.J. Simpson trial.

Radin tried out his machines on the Superbowl of 1996 and even general prime time TV on all four major television stations one evening in February of that year. During the most important moments of the Superbowl game, the machine deviated slightly, but the effect wasn't anywhere near as marked as it was during the O.J. Simpson trial or the Academy Awards. This may have to do with one simple problem with a sports event – the fact that groups of people react differently and passionately to each play, depending on which team they are rooting for. Radin also figured it might have something to do with the number of commercial breaks continually chopping up the game, especially as the advertisements shown during Superbowl have become as popular as the game itself. It was sometimes difficult to distinguish times of high interest from times of low interest and the results showed it.

In his other study of primetime TV, Radin had assumed that both the machines and human observers would peak in the key moments of any show and dribble off at the end, when commercials are usually shown. This is exactly what happened. Although the effect size wasn't enormous, the machine's greater tendency to order peaked just when the audience would have been most involved in the TV shows.

Wagnerians are a fanatical bunch, thought Dieter Vaitl, a colleague of Roger Nelson's, at the Department of Clinical and Physiological Psychology at the University of Giessen. Over the years, the Festspielhaus in Bayreuth, the opera house Wagner had built for himself, had become something of a sacred site to which Wagner aficionados make an annual pilgrimage for the Wagner festival. These were true Wagner fanatics, intimate with every note, every waxing and waning of emotion, happy to sit through 15 hours of the *Ring* cycle. Festspielhaus attendees, in the main, were Wagnerian experts. This, in short, represented the perfect audience for a FieldREG trial.

In 1996, Vaitl, who was very Wagnerian himself, with his sleek pompadour of white hair and his proud demeanor, attended the festival with a FieldREG machine at his side, recording the first cycle of the various operas. He repeated his experiment the following year and the year after that. In total, the REG machine sat through countless hours of Wagner – nine operas, from *Tristan und Isolde* to *Götterdämmerung*. As a whole, over the three years, the trends were consistent, showing an overall change in order in the machine during the most highly emotional scenes or those with the most poignant music, such as choir parts.[12]

In this instance, the PEAR lab couldn't match Vaitl's results. They'd also had a FieldREG machine attend a wide variety of operas and shows in New York City, but the results showed the machines did not react to a significant degree.[13] Obviously, audience attention required a Wagnerian type of intensity to have any affect on the machine. Vaitl concluded that a resonance might be more likely to be created when the audience knows the music well and is tuned into it.

An even more interesting result had come from Radin's other close associate, Professor Dick Bierman in Amsterdam, who had often attempted to replicate his studies. Bierman decided to try out the FieldREG in a home reporting poltergeist-type effects – strange movements or displacement of large objects, usually thought to be caused by ghosts (hence the name, poltergeist, which means 'noisy ghosts'). In some quarters, poltergeists are not believed to be anything more than an intense energy emanating from an individual, often a tempestuous adolescent. In this instance, Bierman installed a REG machine and compared times the family reported a poltergeist effect and the heads-and-tails random output generated by the machine. The same moments the house reported an object flying around, the machine also demonstrated a deviation from chance.[14] It may be that an individual with that type of intensity is creating the poltergeist experience through intense quantum effects in The Field.

Legend has it that the sun always shines on the heads of Princeton alumni, not simply through life but on the day they actually graduate. The local folklore was that even when rain was forecast, it somehow held off until after the commencement exercise was finished. Roger Nelson enjoyed attending the graduation day with his wife every year and had on more than one occasion remarked on the good weather. He now began to wonder whether this was more than simple coincidence. The FieldREG studies had left him with questions about how this type of field consciousness might operate in real life. It occurred to him that the collective wishing of the entire university community for a sunny day might actually have an effect in chasing rain clouds away.

He gathered together all weather reports for the past thirty years and examined what the weather had been like before, during and after the Princeton graduation. Mainly he was looking for the daily rate of precipitation. He also examined the weather of the six towns surrounding Princeton, which were to act as controls.

Nelson's analysis showed some strange effects, as though some collective umbrella surrounded Princeton just on the day its students graduated. In the thirty years, 72 per cent (or nearly three-quarters) of graduation days had been dry, compared with only two-thirds (67 per cent) of days in the surrounding towns. In statistical terms, this meant that Princeton had some magical dry effect around graduation time and was drier than usual, whereas all the surrounding towns were as wet as they should be around that time of year. Even on the one day when there'd been a flood of 2.6 inches of rain in Princeton, curiously the rain had held off until the ceremony had finished.[15]

Nelson's study of the weather in Princeton was only a tiny gauge of whether people could produce a positive effect on their environment. For twenty years, the Transcendental Meditation organization had systematically tested, through dozens and dozens of studies, whether group meditation could reduce violence and discord in the world. It was the contention of the founder of Transcendental Meditation, Maharishi Mahesh Yogi, that individual stress led to world stress and that group calm led to world calm. He'd postulated that if 1 per cent of an area had people practising TM, or the square root of 1 per cent of the population were practising TM-Sidhi, a more advanced and active type of meditation, conflict of any variety – rates of shootings and other crime, drug abuse, even traffic accidents – would go down. The idea of the 'Maharishi' effect was that regularly practicing TM enables you to get in touch with a fundamental field that connects all things – a concept not unlike the Zero Point Field. If enough people were doing it, the coherence would prove infectious among the entire population.

The TM organization had elected to call this 'Super Radiance' because just as superradiance in the brain or in a laser creates coherence and unity, so meditation would have the same effect on society. Special groups of yogic flyers have assembled all over the world, carrying out special 'meditation intensives' targeted at specific areas of conflict. Since 1979 a US Super Radiance group ranging in size from a few hundred to more than 8000 has gathered twice a day at Maharishi International University in Fairfield, Iowa, to attempt to create greater harmony in the world.

Although the TM organization has been ridiculed, largely because of the promotion of the Mararishi's own personal interests, the sheer weight of data is compelling. Many of the studies have been published in impressive journals, such as the *Journal of Conflict Resolution*, the *Journal of Mind and Behavior*, and *Social Indicators Research*, which means that they would have had to meet stringent reviewing procedures. A recent study, the National Demonstration Project in Washington DC, conducted over two months in 1993, showed that when the local Super Radiance group increased to 4000, violent crime, which had been steadily increasing during the first five months of the year, began to fall, to 24 per cent, and continued to drop until the end of the experiment. As soon as the group disbanded, the crime rate rose again. The study demonstrated that the effect couldn't have been due to such variables as weather, the police or any special anti-crime campaign.[16]

Another study of twenty-four US cities showed that whenever a city reached a point where 1 per cent of the population was carrying out regular TM, the crime rate dropped to 24 per cent. In a follow-up study of 48 cities, half of which had a 1 per cent population which meditated, the 1 per cent cities achieved a 22 per cent decrease in crime, compared with an increase of 2 per cent in the control cities, and an 89 per cent reduction in the crime trend, compared with an increase of 53 per cent in the control cities.[17]

The TM organization has even studied whether group meditation could affect world peace. In one 1983 study of a special TM assembly in Israel, which tracked the Arab-Israeli conflict day by day for two months, on days when the number of meditators was high, war deaths in Lebanon fell by 76 per cent, and local crime, traffic accidents and fires all decreased. Once again, confounding influences such as weather, weekends or holidays had been controlled for.[18]

The TM studies, as well as Nelson's FieldREG work, in their own small, preliminary way, offered hope to an alienated and Godless generation. Good might well be able to conquer evil after all. We could create a better community. We had the collective capacity to make the world a better place.

Radin was being a bit facetious when he came up with the idea. He and Nelson had been at Freiburg at a conference in late 1997, and the talk had been about whether they ought to bring some physiological measurements like EEG into studies using REGs. 'Why not look at Gaia's EEG?' Radin remarked at one point.

Nelson immediately pounced on it. As an EEG reads the activity of an individual brain, by attaching electrodes over its surface, so they might be able to take readings of the mind of Gaia, as many people liked to refer to the world. James Lovelock had coined the name, after the Greek goddess of the earth, with his hypothesis that the world is a living entity with its own consciousness.[19] Perhaps they could set up a network of REGs dotted all over the world. The world EEG would be run continuously, taking a constant temperature of the state of the collective mind. When they were searching for a name for it, another colleague of Nelson's came up with 'ElectroGaiaGram', or EGG. Nelson liked the term 'noosphere', coined by Teilhard de Chardin to reflect the idea that the earth was encased in a layer of intelligence. Although Nelson would develop this idea into the

Global Consciousness Project, a project at Princeton but separate from PEAR, EGG was the name that stuck.

If it was true that fields generated by individual consciousnesses can combine during moments of like-mindedness, Nelson wished to see if the collective reaction to the most stirring events of our time would have some sort of common effect on highly sensitive gauges such as REG machines. The O.J. Simpson trial had been a first attempt at this, running machines in different places and comparing the results.

Nelson began with a small group of scientists, who turned on their REG machines in August 1998. He eventually gathered together a network of forty scientists running REGs all over the globe. The project generated a tidal wave of data. Continuous streams of data pouring out of them were sent over the Internet, to be matched with dramatic moments in modern history – the death of John F. Kennedy Jr, and the near impeachment of Bill Clinton; the Paris crash of *Concorde* and the bombing of Yugoslavia; floodings and volcanic eruptions and the New Year's celebrations of Y2K.

Even before EGG started it had its first real test in prototype form, when the world's most beloved princess was suddenly killed in a Paris tunnel. Data recorded before, during and after the Princess of Wales' funeral was compiled and compared with the official schedules of events. During all the public ceremonies for Diana, the machines had veered off their random course, an effect that was 100 to 1 against chance.[20]

However, when Nelson looked at similar data taken during the funeral of Mother Theresa soon after, there had been no untoward effect on the machines. Mother Theresa had been ill and her death had been expected. She was elderly and had lived a full and productive life. Clearly, the tragedy of the young and troubled princess captured the heart of the world, and the REGs had picked it up.[21] American elections and even the Monica Lewinsky scandal didn't seem to stir the world. But New Year's celebrations, major disasters and tragedies sent a shiver through the collective spine that duly showed up on the machines. Not surprisingly, one of the most profound effects was felt during and immediately after the September 11 terrorist attacks on the World Trade Center.[22]

These initial results left Nelson and Radin with many tantalizing questions. If there was such a thing as a world mind, perhaps little flashes of inspiration in it could account for the most monstrous and magnificent moments in human history, or maybe negative consciousness was also like a germ that could infect people and take hold. Germany had been depressed in every sense after the First World War. Could this dispiritedness have affected the Germans on a quantum level, making it easier for Hitler, that most intoxicating of speakers, to create a kind of negative collective, which fed on itself and condoned the grossest of evils? Had a collective consciousness been responsible for the Spanish inquisition? The Salem witchcraft trials? Did collective evil also create coherence?

And what of man's greatest achievements? Could a sudden gust of inspiration occur in the world mind? Could some coalescence of energy be responsible for the flowering of art or higher consciousness in a certain age? For the ancient Greeks? The Renaissance? Was creativity also infectious, accounting for the explosive creativity in Vienna in the 1790s and the burgeoning of British pop music in the 1960s? The Zero Point Field provided a likely explanation for certain unexplained physical synchronicities – such as the scientifically verified coming together of menstrual cycles among women in close proximity.[23] Could it also account for emotional and intellectual synchronicity in the world?

It was the first inkling that group consciousness, working through a medium such as the Zero Point Field, acted as the universal organizing factor in the cosmos. But so far, with the technology to hand, Nelson had only the first glimmers of evidence, a tiny deviation from random activity. All he could do thus far was measure a single pebble or at best a handful of sand – the quantum effect of an individual or a small group on the world. One day, he might have the capacity to measure the effect of the entire beach, for that was the ultimate point. The beach should only be measured in its entirety. The sand of the entire shore is indivisible.

Twenty-five years after Edgar Mitchell had experienced collective consciousness viscerally, scientists were beginning to prove it in a laboratory.[24]

CHAPTER TWELVE

The Zero Point Age

IN A DRAB LITTLE corner classroom at the UK's University of Sussex on a frosty day in January 2001, a group of sixty scientists from ten countries had crowded together to try to work out exactly how they were going to fly 20 trillion miles into deep space. NASA had had a few Breakthrough Propulsion Physics workshops in America and this was to be the international equivalent: one of the first independent workshops ever held on propulsion. Indeed, it had attracted an impressive audience of physicists from the British government, a NASA marshal, various astrophysicists from the French Laboratoire D'Astrophysics Marseilles and the French Laboratory of Gravitation, Relativity and Cosmology, professors from American and European universities, and some fifteen representatives of private industry. This was just a seed meeting, not a true scientific conference, mainly to start the ball rolling – a precursor to the international conference to be held December 2001. Nevertheless, there was an unmistakable air of expectancy around the room, tacit acknowledgment that each person present was perched on the very frontier of scientific knowledge and might even be witness to the dawning of a new age. Graham Ennis, the conference organizer, had lured representatives from most of the major British newspapers and science magazines by dangling before them the prediction that in five years' time we'd be building our own small rockets with WARP drives to keep satellites in their correct positions.

However distinguished the audience, the greatest deference was reserved for Dr Hal Puthoff, by now in his early sixties, a bit thinner but still with his thatch of greying hair, who'd spent nearly thirty years trying to determine whether you could harness the space between the stars. To a handful of the younger members of the audience, Hal had become something of a cult figure. A young British government physicist called Richard Obousy had stumbled across Hal's Zero Point Field papers during his university studies, and been thunderstruck by their implications, so much so that they'd influenced the course of his own career.[1] And now he was faced with the prospect of both meeting the great man and preceding him on the podium with a small introductory talk on manipulating the vacuum

– a warm-up act to the day's main attraction.

To any outside observation, this was something more than a frivolous exercise, a batch of technocrats playing at constructing the ultimate techno-toy. It was clear to every scientist in the room that the planet had, at most, fifty years of fossil fuel left and humans were facing a climate crisis as the greenhouse effect slowly turned our world into a gas chamber. Looking for new sources of energy wasn't just necessary to power spaceships. It was also vital to power earth and maintain it intact for the next generation.

Experiments making use of the most outlandish of new ideas in physics had been going on covertly for thirty years. Rumors abounded about secret testing sites at places like Los Alamos with billion-dollar 'black' budgets that NASA or the American military continued to hotly deny. Even British Aerospace had launched its own secret program – code-named Project Greenglow – to study the possibility of turning off gravity.[2]

Loads of other possibilities, all resting on solid, proven physics, might provide for new methods of space-flight propulsion, said Ennis, who was presiding over the first day. You could: control inertia, so that you could move large things such as spacecraft with small forces; use one of a number of nuclear fusion techniques, which would require tremendous pressure and temperature; employ a radioactive fission reactor, as the Russians had done; use tethers, which would extract electrostatic energy; employ matter–antimatter effects, where the reaction of matter meeting its opposite number creates energy; change electromagnetic fields; or rotate superconductors. At a NASA congress in Albuquerque, New Mexico, they'd been exploring the possibility of a spaceship creating its own wormhole, much as Carl Sagan had imagined in *Contact*.[3] A number of private companies, including Lockheed Martin, were enthusiastic and had lent their support. This could have all sorts of practical everyday applications on earth. Imagine, for instance, if you could turn off gravity and levitate patients. You could make bedsores a thing of the past.

Or you could try something even more outlandish. You could try to extract your energy from the nothingness of space itself. The 'ZPF', scientists agreed, represented one of the best possible scenarios – a 'cosmic free lunch', as Graham Ennis liked to put it, an endless supply of something from nothing. After physicist Robert Forward of Hughes Research Laboratory in Malibu, California, wrote a paper about it, theorizing how you might conduct experiments,[4] physicists were beginning to believe that it may be possible to get to it and, more importantly, get energy out of it.

During his talk the following day, Hal Puthoff explained that, in quantum mechanical terms, if you were going to attempt to extract energy from The Field, you'd have several choices. You'd need to decouple from gravity, reduce inertia or generate enough energy from the vacuum to overcome both. The US Air Force had first recommended that Forward do his study to measure the Casimir force, the quantum force between two metal plates caused by partially shielding the space between them from zero-point fluctuations in the vacuum and so unbalancing the zero-point energy radiations. Forward, an expert in gravitational theory, was given the assignment by the Propulsion Directorate of the Phillips Laboratory at Edwards Air Force Base, which has the task of launching research into twenty-first-century space propulsion.

They had proof that vacuum fluctuations could be altered using technology. However, Casimir forces are unimaginably small – a pressure of just one hundred-millionth of an atmosphere on plates held a thousandth of a millimeter apart.[5] Bernie Haisch and Daniel Cole published a paper theorizing that if you built a vacuum engine of an enormous number of such colliding plates, each would generate heat when they finally come into contact and give you power. The problem is that each plate creates, at most, a half of a microwatt's worth of energy – 'not much to write home about', said Puthoff.[6] You'd need tiny systems running at a very high rate for it to work on any level.

Forward thought that it was possible to do an experiment on altering inertia by making changes in the vacuum. He recommended four such experiments to be carried out to test this concept.[7] Scientists working in quantum electrodynamics had already shown that these vacuum fluctuations could be controlled once you manipulated the spontaneous emission rates of atoms. It was Puthoff's view that electrons get their energy to whiz around the nucleus of an atom without slowing down because they are tapping quantum fluctuations of empty space. If we could manipulate that field, he said, we could destabilize atoms and extract the power from them.[8]

It was theoretically possible to extract energy from the Zero Point Field; even in nature scientists had conjectured that this was exactly what was happening when cosmic rays 'power up' or energy is released by supernovas and gamma-ray bursters. There were other ideas, such as the spectacular conversion of sound into light waves, or sonoluminescence, where water, bombarded with intense sound waves, creates air bubbles which rapidly contract and collapse in a flash of light. The theory in some quarters was that this phenomenon was caused by zero-point energy inside the bubbles, which, once the bubbles shrank, converted into light. But Puthoff had already tried all these ideas in turn and felt they held little promise.

The US Air Force had also been exploring the idea of cosmic rays driven by zero-point energy, where protons could be accelerated in a cryogenically cooled, collision-free vacuum trap – a chamber that had been cooled as close as possible to absolute zero. This would give you about the emptiest space possible to attempt to extract energy from vacuum fluctuations of protons once they started to go faster. Another idea was downshifting the more energetic high-frequency parts of zero-point energy through the use of specially created antennae.

In his own laboratory, Puthoff had been playing around with a method that would involve perturbing ground states of atoms or molecules. According to his own theories, these were simply equilibrium states involving the dynamic radiation/absorption exchange with the Zero Point Field. So if you employed some sort of Casimir cavity, the atoms or molecules might undergo energy shifts that would alter excitations involving the ground states. He'd already begun experiments at a synchrotron facility, a place with a special subatomic accelerator, to try this, but had so far met with failure.[9]

Then Hal thought of turning the whole project inside out, following up on a notion first mooted by general relativity theorist Miguel Alcubierre of the University of Wales. Alcubierre had tried to determine whether WARP drives, as described in *Star Trek*, really were possible.[10] Suppose you ignore quantum theory and look upon this as a problem of general relativity. Instead of invoking Niels Bohr, you invoke Albert Einstein. What if you tried modifying the space-time metric? If you use the curved space-time of Einstein, you treat the vacuum as a medium that could be polarized. You do a little 'vacuum engineering,' as Nobel prize laureate Tsung-Dao Lee called it.[11] Under this interpretation, the bending of a light ray, say, near a massive body, is caused by a variation in the refractive index of the vacuum near that mass. The propagation of light defines the space-time metric. What you might be able to do is decrease the refractive index of the Zero Point Field, which would then increase the speed of light. If you modify space-time to an extreme degree, the speed of light is greatly increased. Mass then decreases and energy-bond strength increases –

features that theoretically would make interstellar travel possible.

What you do is to distort and expand space-time behind the spaceship, contract space-time in front of it, and then surf along on it faster than the speed of light. In other words, you restructure general relativity as an engineer would. If you could successfully do this, you could make a spaceship travel at ten times the speed of light, which would be apparent to people on earth but not to the astronauts inside. You'd finally have yourself a *Star Trek* WARP drive.

What you are doing by such 'metric engineering', as Hal termed it, is getting space-time to push you away from the earth and toward your destination. This is possible by creating large-scale Casimir-like forces. Another possible type of metric engineering, which also requires using Casimir forces, is traveling through wormholes – 'cosmic subways'[12], as Hal referred to them, which connect you to distant parts of the universe, as was imagined in *Contact*.

'But how close were we to doing any of this?', the audience asked. Hal coughed to clear his throat, his characteristic tic. It might take twenty years to do it, he replied laconically. Or it might take that same amount of time just to decide that it was not possible to get to it. You probably weren't looking at major space travel in his lifetime, although he still held out hope of extracting energy for earthbound fuel before he died.

The first international propulsion workshop was an undoubted success, a good meeting place for physicists who'd been working away on their own at problems of energy and thrust that might take half a century to see the light of day. It was evident to everyone that they were at the beginning of an exploration that would one day, as Arthur C. Clarke had put it, make today's current efforts at venturing beyond our atmosphere look like nineteenth-century attempts to conquer flight with a hot-air balloon.[13] But in different parts of the world, many of Puthoff's old colleagues, also now in their sixties, were working away without fanfare on more earthbound activities that were every bit as revolutionary, all predicated on the idea that all communication in the universe exists as a pulsating frequency and The Field provides the basis for everything to communicate with everything else.

In Paris, The DigiBio team, still in its Portakabin, had by now perfected the art of capturing, copying and transferring the electromagnetic signals from cells. Since 1997, Benveniste and his DigiBio colleagues have filed three patents on diverse applications. For Benveniste the biologist, the applications, naturally enough, were medical. He believed his discovery could open the way for an entirely new digital biology and medicine, which would replace the current clumsy hit-and-miss method of taking drugs.

It occurred to him that if you don't need the molecule itself, but only its signal, then you don't need to take drugs, do biopsies or test for toxic substances or pathogens such as parasites and bacteria with physical sampling. As he'd already shown in one study, you could use frequency signaling to detect the bacteria *E. coli*.[14] It's known that latex particles sensitized to a certain antibody will cluster in the presence of *E. coli* K1. By recording the signal for *E. coli*, another bacteria and also control substances, and then applying them to the latex particles, Benveniste found that the *E. coli* produced the largest clusters of any of the frequencies. Before long, his team's record for detecting the *E. coli* signal became virtually perfect.

Using digital recording, we could uncover those pathogens like prions, which have no reliable means of detection, and no longer waste precious laboratory resources in determining whether antigens are present in the body and whether the body has mounted antibodies to them. It also may mean that when we are ill, we may not need to take drugs. We could get rid of unwanted parasites or bacteria just by playing an unfriendly frequency. We could use electromagnetic means of detecting dangerous microorganisms in our agriculture or use them to find out whether foods have been genetically modified. If we could come up with the right frequencies, we wouldn't have to use dangerous pesticides but could just kill bugs with electromagnetic signals. You wouldn't even have to do all this detection work in person. Virtually all the test samples could be emailed and carried out remotely.

In America, the AND Corporation, a company with offices in New York, Toronto and Copenhagen, was working away at artificial intelligence based upon the ideas of Karl Pribram and Walter Schempp about how the brain works. Its proprietary system, called Holographic Neural Technology (Hnet), for which it now has a worldwide patent, used principles of holography and wave encoding for computers to learn tens of thousands of stimulus-response memories in less than a minute and to respond to tens of thousands of these patterns in less than a second. In AND's view, its system was an artificial replica of how the brain works. Single neuron cells with just a few synapses were capable of learning memories instantly.

Millions of these memories could be superimposed. The model demonstrates how these cells can memorize abstraction – a concept, say, or a human face. AND had ambitious plans for its technology. It was planning to set up Strategic Business Units, in different specialties, which, if developed properly, might transform the information processing of virtually any industry.

Fritz-Albert Popp and his team of IIB scientists were beginning to test biophoton-emission detection as a means of determining whether food was fresh. His experiments and the theoretical approach behind them were gaining acceptance among the scientific community.

Dean Radin put some of his studies up on the Internet for visitors to participate in, and engaged in giant computerized experiments. Braud and Targ carried on with more studies of human intention and healing. Brenda Dunne and Bob Jahn carried on adding to their mountain of data. Roger Nelson, with his Global Project, continued to measure small tremors on the collective cosmic seismograph.

Edgar Mitchell presented the keynote address of CASYS 1999, an annual mathematical conference in Liège, Belgium, sponsored by the Society for the Study of Anticipatory Systems, which incorporated his synthesis of theories of quantum holography and human consciousness. The discovery of the presence of quantum resonance in living things and the ability of the Zero Point Field to encode information and provide instantaneous communication represented no less than the Rosetta Stone of human consciousness, he said.[15] All the different strands he'd been investigating for thirty years were finally beginning to come together.

At that same conference, he and Pribram were honored together for their exploration of outer space and inner space – Pribram for his scientific work on the holographic brain, and Mitchell for outstanding scientific work on noetic sciences. That same year, Pribram received the Dagmar and Václav Havel prize for bringing together the sciences and humanities.

Hal Puthoff sat on the unofficial subcommittee of NASA's Breakthrough Propulsion Program: the Advanced Deep Space Transport (ADST) Group – a group of people, he said, who are on the 'frontier of the frontier'.[16]

In his capacity as director of the Institute for Advanced Studies, Hal operated as a clearinghouse for inventors or companies who think they've developed a gadget of any sort that taps into the Zero Point Field. Hal would put each one to the ultimate test – it must show that more energy is coming out of the gadget than going in. Thus far, every one of the thirty devices tested by him has failed. But he is still optimistic, as only a frontier scientist can be.[17]

In terms of the real import of their discoveries these practical uses represented only a bit of technological froth. All of them – Robert Jahn and Hal Puthoff, Fritz-Albert Popp and Karl Pribram – were philosophers as well as scientists, and on rare occasions when they weren't busy pressing on with their experimentation, it had occurred to them that they had dug deep and come up with something profound – possibly even a new science. They had the beginnings of an answer to much of what had remained missing in quantum physics. Peter Milonni at Los Alamos's NASA facilities had speculated that if the fathers of quantum theory had used classical physics with the Zero Point Field, the scientific community would have been far more satisfied with the result than they were by the many unanswerables of quantum physics.[18] There are those who believe quantum theory will one day be replaced by a modified classical theory which takes into account the Zero Point Field. The work of these scientists may take the word 'quantum' out of quantum physics and create a unified physics of the world, large and small.

Each scientist had taken his own incredible voyage of discovery. As young scientists with promising credentials, each had begun his career holding certain tenets sacred – the ideas and received wisdom of their peers:

> The human being is a survival machine largely powered by chemicals and genetic coding.
> The brain is a discrete organ and the home of consciousness, which is also largely driven by chemistry – the communication of cells and the coding of DNA.
> Man is essentially isolated from his world, and his mind is isolated from his body.
> Time and space are finite, universal orders.
> Nothing travels faster than the speed of light.

Each of them had chanced upon an anomaly in this thinking and had the courage and the independence to pursue that line of inquiry. One by one, through painstaking experiment and trial and error, each had eventually come to the position that every one of these tenets – bedrocks of physics and biology – were probably wrong:

> The communication of the world did not occur in the visible realm of Newton, but in the subatomic world of Werner Heisenberg.
> Cells and DNA communicated through frequencies.
> The brain perceived and made its own record of the world in pulsating waves.
> A substructure underpins the universe that is essentially a recording medium of everything, providing a means for everything to communicate with everything else.
> People are indivisible from their environment. Living consciousness is not an isolated entity. It increases order in the rest of the world. The consciousness of human beings has incredible powers, to heal ourselves, to heal the world – in a sense, to make it as we wish it to be.

Every day in their laboratories, these scientists caught a tiny glimmer of the possibilities suggested by their discoveries. They'd found that we were something far more impressive than evolutionary happenstance or genetic survival machines. Their work suggested a decentralized but unified intelligence that was far grander and more exquisite than Darwin or Newton had imagined, a process that was not random or chaotic, but intelligent and purposeful. They'd discovered that in the dynamic flow of life, order triumphed.

These are discoveries that may change the lives of future generations in many practical ways, in fuel-less travel and instant levitation; but in terms of understanding the furthest reaches of human potential, their work suggested something far more profound. In the past, individuals had accidentally evidenced some ability – a premonition, a 'past life', a clairvoyant image, a gift for healing – which quickly was dismissed as a freak of nature or a confidence trick. The work of these scientists suggested that this was a capacity neither abnormal nor rare, but present in every human being. Their work hinted at human abilities beyond what we'd ever dreamed possible. We were far more than we realized. If we could understand this potential scientifically, we might then learn how to systematically tap into it. This would vastly improve every area of our lives, from communication and self-knowledge to our interaction with our material world. Science would no longer reduce us to our lowest common denominator. It would help us take a final evolutionary step in our own history by at last understanding ourselves in all of our potential.

These experiments had helped to validate alternative medicine, which has been shown to work empirically but has never been understood. If we could finally work out the science of medicine that treats human energy levels and the exact nature of the 'energy' that was being treated, the possibilities for improved health were unimaginable.

These were also discoveries which scientifically verified the ancient wisdom and folklore of traditional cultures. Their theories offered scientific validation of many of the myths and religions humans have believed in since the beginning of time, but have hitherto only had faith to rely on. All they'd done was to provide a scientific framework for what the wisest among us already knew.

Traditional Australian Aborigines believe, as do many other 'primitive' cultures, that rocks, stones and mountains are alive and that we 'sing' the world into being – that we are creating as we name things. The discoveries of Braud and Jahn showed that this was more than superstition. It was just as the Achuar and the Huaorani Indians believe. On our deepest level, we do share our dreams.

The coming scientific revolution heralded the end of dualism in every sense. Far from destroying God, science for the first time was proving His existence – by demonstrating that a higher, collective consciousness was out there. There need no longer be two truths, the truth of science and the truth of religion. There could be one unified vision of the world.

This revolution in scientific thinking also promised to give us back a sense of optimism, something that has been stripped out of our sense of ourselves with the arid vision of twentieth-century philosophy, largely derived from the views espoused by science. We were not isolated beings living our desperate lives on a lonely planet in an indifferent universe. We never were alone. We were always part of a larger whole. We were and always had been at the center of things. Things did not fall apart. The center did hold and it was we who were doing the holding.

We had far more power than we realized, to heal ourselves, our loved ones, even our communities. Each of us had the ability – and together a great collective power – to improve our lot in life. Our life, in every sense, was in our hands.

These were bold insights and discoveries but very few had heard them. For thirty years, these pioneers had presented their findings at small mathematical conferences or the annual meetings of tiny scientific bodies created to promote a dialogue on frontier science. They knew and admired each other's work and were acknowledged at these small gatherings of their peers. Most of the scientists had been young men when they made their discoveries, and before they embarked on what turned out to be lifelong detours they had been highly respected, even revered. Now they were approaching retirement age, and among the wider scientific community most of their work still had never seen the light of day. They were all Christopher Columbus and nobody believed what they'd returned to tell. The bulk of the scientific community ignored them, continuing to grip tightly to the notion that the earth was flat.

The space-propulsion activities had been the only acceptable face of the Zero Point Field. Despite their rigorous scientific protocols, nobody in the orthodox community was taking any other discoveries of theirs seriously. Some, like Benveniste, had merely been marginalized. For many years, Edgar Mitchell, now 71, depended on lectures about his exploits in outer space to fund his research into consciousness. Every so often Robert Jahn would submit a paper with unimpeachable statistical evidence to an engineering journal, and they would dismiss it out of hand. Not for the science, but for its shattering implications about the current scientific world view.

Nevertheless, Jahn and Puthoff and the other scientists all knew what they had. Each carried on with the stubborn blinkered confidence of the true inventor. The old way was simply one more hot-air balloon. Resistance was the way it had always been in science. New ideas were always considered heretical. Their evidence might well change the world forever. There were many areas to be refined, other paths to go down. Many might turn out to be detours or even dead ends, but the first tentative inquiries had been made. It was a start, a first step, the way all real science started.

VISIT OUR WEBSITE . . .

If you enjoyed *The Field,* and would like to find out more about the latest discoveries, or how to live 'in the Field,' visit our website at:

www.thefieldonline.com

NOTES

Unless otherwise indicated, all information about the scientists and the details of their discoveries was culled from multiple telephone interviews conducted between 1998–2001.

ACKNOWLEDGMENTS

1 D. Reilly, 'Is evidence for homeopathy reproducible?' *The Lancet*, 1994; 344: 1601–6.

PROLOGUE: THE COMING REVOLUTION

1 M. Capek, *The Philosophical Impact of Contemporary Physics* (Princeton, New Jersey: Van Nostrand, 1961): 319, as quoted in F. Capra, *The Tao of Physics* (London: Flamingo, 1992).
2 D. Zohar, *The Quantum Self* (London: Flamingo, 1991): 2; Danah Zohar provides an excellent summary of the philosophical history of science before and after Newton and Descartes.
3 I am indebted to Brenda Dunne, manager of the PEAR laboratory in Princeton, for first enlightening me about the philosophical interests of the quantum theorists. See also W. Heisenberg, *Physics and Philosophy* (Harmondsworth: Penguin, 2000), N. Bohr, *Atomic Physics and Human Knowledge* (New York: John Wiley & Sons, 1958), and R. Jahn and B. Dunne, *Margins of Reality: The Role of Consciousness in the Physical World* (New York: Harvest/Harcourt Brace Jovanovich, 1987): 58–9.
4 Interview with Robert Jahn and Brenda Dunne, Amsterdam, October 19, 2000.
5 Indeed, in determining which of the scientists merited inclusion, I have had to make certain arbitrary selections. I chose American anesthesiologist Stuart Hameroff and his work on human consciousness, when I could as easily have chosen Oxford professor Roger Penrose. Only for reasons of space have I omitted pioneers into electromagnetic cell communications like Cyril Smith.

CHAPTER ONE: LIGHT IN THE DARKNESS

1 For an account of Dr Mitchell's voyage, I have relied on E. Mitchell, *The Way of the Explorer: An Apollo Astronaut's Journey Through the Material and Mystical Worlds* (G.P. Putnam, 1996): 47-56; M. Light, *Full Moon* (London: Jonathan Cape, 1999); a visit to an exhibition of lunar photographs (London: Tate Gallery, November 1999); personal interviews with Dr Mitchell (summer and autumn 1999); T. Wolfe, *The*

Right Stuff (London: Jonathan Cape, 1980); and A. Chaikin, *A Man on the Moon* (Harmondsworth: Penguin, 1994): 355-79.

2 Mitchell, *Way of the Explorer*: 61. Dr Mitchell's results were published in the *Journal of Parapsychology*, June 1971.

3 Francis Crick likened the brain to a TV set, as quoted in D. Loye, *An Arrow Through Chaos* (Rochester, Vt: Park Street Press, 2000): 91.

4 Nonlocality was considered to be proven by experiments conducted by Alain Aspect and his colleagues in Paris in 1982.

5 M. Schiff, *The Memory of Water: Homeopathy and the Battle of Ideas in the New Science* (Thorsons, 1995).

CHAPTER TWO: THE SEA OF LIGHT

Detail on the U.S. oil crisis was compiled from articles appearing in the *London Times*, Nov. 26–Dec. 1, 1973.

1 H. Puthoff, 'Everything for nothing', *New Scientist*, July 28, 1990: 52–5.

2 J. D. Barrow, *The Book of Nothing* (London, Jonathan Cape, 2000): 216.

3 A simple equation showing energy for harmonic oscillators would be represented as $H = \Sigma_i \hbar\Omega_i(n_i + 1/2)$. The 1/2 stood for the zero-point energy. When renormalizing, scientists would just drop the 1/2. Communication with Hal Puthoff, December 7, 2000.

4 The Zero Point Field is included in stochastic electrodynamics. But in ordinary classical physics, it is usually 'renormalized' away.

5 T. Boyer, 'Deviation of the black-body radiation spectrum without quantum physics', *Physical Review*, 1969; 182: 1374–83.

6 Interviews with Richard Obousy, January 2001.

7 R. Sheldrake, *Seven Experiments That Could Change the World* (London: Fourth Estate, 1994): 75–6.

8 R. O. Becker and G. Selden, *The Body Electric* (Quill, 1985): 81.

9 A. Michelson and E. Morley, *American Journal of Science*, 1887, series 3; 34: 333–45, cited in Barrow, *Book of Nothing*: 143–4.

10 Quoted in F. Capra, *The Tao of Physics* (London: Flamingo, 1976).

11 E. Laszlo, *The Interconnected Universe: Conceptual Foundations of Transdisciplinary Unified Theory* (Singapore: World Scientific, 1995).

12 A. C. Clarke, 'When will the real space age begin?', *Ad Astra*, May/June 1996: 13–5.

13 B. Haisch, 'Brilliant disguise: light, matter and the Zero Point Field', *Science and Spirit*, 1999; 10: 30–1. Elsewhere, Dr Haisch has made numerous interesting speculations about the connection between Creation and the Zero Point Field and refers to the ZPF as a 'sea of light'. For the agnostic, the theory is that the random background fluctuations of the vacuum are residual energy left over from the Big Bang. See H. Puthoff, *New Scientist*, July 28, 1990: 52. Particle physicists theorize that the universe was created as a false vacuum, with more energy than it ought to have had. When that energy decayed, it produced an ordinary quantum vacuum, which led to the Big Bang and produced all the energy for mass in the universe. See H. E. Puthoff, 'The energetic vacuum: implications for energy research', *Speculations in Science and Technology*, 1990; 13: 247–57.

14 H. Puthoff, 'Ground state of hydrogen as a zero-point-fluctuation-determined state,' *Physical Review D*; 1987, 35: 3266–70.

15 Interview with Bernhard Haisch, California, October 29, 1999.

16 J. Gribbin, *Q is for Quantum: Particle Physics from A to Z* (Phoenix, 1999): 66;

H. Puthoff, 'Everything for nothing': 52.

17 Puthoff, 'Ground state of hydrogen'. Also, conversations with Hal Puthoff, July 20, and August 4, 2000, and Benhard Haisch, October 26, 1999.

18 H. E. Puthoff 'Source of vacuum electromagnetic zero-point energy', *Physical Review A*, 1989: 40: 4857–62; also reply to comment, 1991; 44: 3385–6.

19 H. Puthoff, 'Where does the zero-point energy come from?', *New Scientist*, December 2, 1989: 36.

20 H. Puthoff, 'The energetic vacuum: implications for energy research, *Speculations in Science and Technology*, 1990; 13: 247-57.

21 Ibid.

22 In the Zero Point Field, Puthoff also found an explanation for the cosmological coincidence first discovered by British physicist Paul Dirac. This showed that the average density of matter – the average pull between an electron and a proton – has a close relationship to the size of the universe – measured by the ratio of the size of the universe to the size of an electron. Puthoff found that this was just related to the density of Zero Point Field energy. See *New Scientist*, December 2, 1989.

23 Various conversations with Hal Puthoff, 2000 and 2001; also H. Puthoff, 'On the relationship of quantum energy research to the role of metaphysical processes in the physical world', www.meta-list.org.

24 Puthoff, 'Everything for nothing'.

25 S. Adler (in a selection of short articles dedicated to the work of Andrei Sakharov), 'A key to understanding gravity', *New Scientist*, April 30, 1981: 277–8.

26 B. Haisch, A. Rueda and H. E. Puthoff, 'Beyond $E=mc^2$: A first glimpse of a universe without mass', *The Sciences*, November/December 1994: 26–31.

27 Puthoff, 'Everything for nothing'.

28 H. E. Puthoff, 'Gravity as a zero-point-fluctuation force,' *Physical Review A*, 1989; 39(5): 2333–42; also 'Comment', *Physical Review A*, 1993; 47(4): 3454–5.

29 Ibid.

30 Interview with Hal Puthoff, April 8, 2000.

31 Energy Conversion using High Charge Density, US Patent no. 5,018,180.

32 Interview with Bernhard Haisch, California, October 26, 1999.

33 Robert Matthews, 'Inertia: does empty space put up the resistance?' *Science*, 1994; 263: 613. This property of the vacuum was also tested by Stanford Linear Accelerator Center.

34 B. Haisch, A. Rueda and H. E. Puthoff, 'Inertia as a zero-point-field Lorentz force', *Physical Review A*, 1994; 49(2): 678–94.

35 B. Haisch, A. Rueda and H. E. Puthoff, paper presented at AIAA 98-3143, Advances ASME/SAE/ASEE Joint Propulsion Conference & Exhibit, July 13–15, 1998, Cleveland, Ohio; also B. Haisch, 'Brilliant Disguise.'

36 Haisch *et al.*, 'Beyond $E=mc^2$'.

37 A. C. Clarke, *3001: The Final Odyssey* (HarperCollins, 1997): 258.

38 Ibid.

39 Ibid.: 258–9.

40 Clarke, 'When will the real space age begin?': 15.

41 A. Rueda, B. Haisch and D. C. Cole, 'Vacuum zero-point field pressure instability in astrophysical plasmas and the formation of cosmic voids', *Astrophysical Journal*, 1995; 445: 7–16.

42 R. Matthews, 'Inertia'.
43 D. C. Cole and H. E. Puthoff, 'Extracting energy and heat from the vacuum', *Physical Review E*, 1993; 48(2): 1562–5.
44 Interview with Bernhard Haisch, California, October 29, 1999.
45 Interviews with Hal Puthoff, July and August 2000; also H. Puthoff, 'On the relationship of quantum energy'. I have deliberately used a few of Puthoff's phrases from his unpublished article to indicate his thinking at the time.
46 Clarke, 'When will the real space age begin?'.

CHAPTER THREE: BEINGS OF LIGHT

1 F. A. Popp, 'MO-Rechnungen an 3,4-Benzpyren und 1,2-Benzpyren legen ein Modell zur Deutung der chemischen Karzinogenese nahe', *Zeitschrift für Naturforschung*, 1972; 27b: 731; F. A. Popp, 'Einige Möglichkeiten für Biosignale zur Steuerung des Zellwachstums', *Archiv für Geschwulstforschung*, 1974; 44: 295–306.
2 B. Ruth and F. A. Popp, 'Experimentelle Untersuchungen zur ultraschwachen Photonememission biologisher Systeme', *Zeitschrift für Naturforschung*, 1976; 31c: 741–5.
3 M. Rattemeyer, F. A. Popp and W. Nagl, *Naturwissenschaften*, 1981; 11: 572–3.
4 R. Dawkins, *The Selfish Gene*, 2nd edn (Oxford: Oxford University Press, 1989): 22.
5 Ibid.: preface, 2; see also R. Sheldrake, *The Presence of the Past* (London: Collins, 1988): 83–5.
6 Dawkins, *Selfish Gene*: 23.
7 Ibid.: 23; 'This, at the present time in molecular biology, is the learned soundscreen of language behind which is hidden the ignorance, for want of a better explanation.'
8 Telephone interview with Fritz-Albert Popp, January 29, 2001.
9 R. Sheldrake, *A New Science of Life* (London: Paladin, 1987): 23–5.
10 R. Sheldrake, *A New Science of Life: The Hypothesis of Formative Causation* (London: Blond and Briggs, 1981); Sheldrake, *Presence of the Past*.
11 Sheldrake has expressed the view that nonlocality in quantum physics might ultimately explain some of his theories. See Sheldrake website: www.sheldrake.org.
12 See H. Reiter und D. Gabor, *Zellteilung und Strahlung. Sonderheft der Wissenschaftlichen Veroffentlichungen aus dem Siemens-Konzern* (Berlin: Springer, 1928).
13 R. Gerber, *Vibrational Medicine* (Santa Fe: Bear and Company, 1988): 62.
14 H. Burr, *The Fields of Life* (New York: Ballantine, 1972).
15 R. O. Becker and G. Selden, *The Body Electric: Electromagnetism and the Foundation of Life* (Quill, 1985): 83.
16 Experiments by Lund, Marsh and Beams are recounted in Becker and Selden, *The Body Electric*: 82–5.
17 Becker and Selden, *Body Electric*: 73–4.
18 H. Fröhlich, 'Long-range coherence and energy storage in biological systems', *International Journal of Quantum Chemistry*, 1968; 2: 641–9.
19 H. Fröhlich, 'Evidence for Bose condensation-like excitation of coherent modes in biological systems', *Physics Letters*, 1975, 51A: 21; see also D. Zohar, *The Quantum Self* (London: Flamingo, 1991): 65.
20 R. Nobili, 'Schrödinger wave holography in brain cortex', *Physical Review A*, 1985; 32: 3618–26; R. Nobili, 'Ionic waves in animal tissues', *Physical Review A*, 1987; 35: 1901–22.
21 Becker and Selden, *The Body Electric*: 92–3; also R. Gerber, *Vibrational Medicine*: 98; M. Schiff, *The Memory of Water*: 12. More recently, another Italian, Ezio In-

sinna, proposed that centrioles, the little cartwheel structures holding cell structure in place, are virtually 'immortal' oscillators, or wave generators. In an embryo, these waves will be set in motion by the father's genes when they first unite with the mother's genes, and thereafter continue pulsing through the life of the organism. At the first stage of an embryo's development, they might begin at a certain frequency to affect cell shape and metabolism, and then change the frequency as the organism matures. Correspondence with E. Insinna, November 5, 1998. See E. Insinna, 'Synchronicity and coherent excitations in microtubules', *Nanobiology*, 1992; 1: 191–208; 'ciliated cell electrodynamics: from cilia and flagella to ciliated sensory systems', in A. Malhotra, ed., *Advances in Structural Biology*, Stamford, Connecticut: JAI Press, 1999:5. T. Y. Tsong has also written about the electromagnetic language of cells: T. Y. Tsong, 'Deciphering the language of cells', *Trends in Biochemical Sciences*, 1989; 14: 89–92.

22 F. A. Popp, Qiao Gu and Ke-Hsueh Li, 'Biophoton emission: experimental background and theoretical approaches', *Modern Physics Letters B*, 1994; 8(21/22): 1269– 96; also, F. A. Popp, 'Biophotonics: a powerful tool for investigating and understanding life', in H. P. Dürr, F. A. Popp and W. Schommers (eds), *What is Life?* (Singapore: World Scientific), in press.

23 S. Cohen and F. A. Popp, 'Biophoton emission of the human body', *Journal of Photochemistry and Photobiology B: Biology*, 1997; 40: 187–9.

24 Interviews with Fritz-Albert Popp, Coventry and telephone, March 2001.

25 F. A. Popp and Jiin-Ju Chang, 'Mechanism of interaction between electromagnetic fields and living systems', *Science in China (Series C)*, 2000; 43: 507–18.

26 Biologist Rupert Sheldrake has recently made a study of the special abilities of animals. His own studies have demonstrated that termite colonies will make columns and then bend them toward each other until the ends of the new columns meet in an arch, according to some master plan beyond all usual communication. One of the best experiments testing this ability was carried out by South African naturalist Eugene Marais, who placed a steel plate in a termite mound. Despite the height and width of the plate, the termites would build an arch or tower on each side of the plate so similar that when the steel plate was withdrawn, the two halves matched perfectly. Marais (and later Sheldrake) concluded that the termites operate according to an organizing energy field far more advanced than any sensory communication, particularly since many forms would not be able to penetrate the steel plate. Sheldrake has amassed a database of 2,700 case histories of pets and apparent telepathic behavior, and a number of surveys with pet owners. More than 200 studies concern the telepathic abilities of JayTee, a mixed-breed terrier in the north of England, who will go to the window and wait for his owner, Pamela Smart, in telepathic anticipation of her arrival, even if she sets off for home at unusual times and in strange vehicles. See R. Sheldrake, *Seven Experiments That Could Change the World: A Do-It-Yourself Guide to Revolutionary Science* (Fourth Estate, 1994): 68–86, and *Dogs That Know When Their Owners Are Coming Home and Other Unexplained Powers of Animals* (Hutchinson, 1999).

27 Interview with Fritz-Albert Popp, Coventry, March 21, 2001.

28 J. Hyvarien and M. Karlssohn, 'Low-resistance skin points that may coincide with acupuncture loci', *Medical Biology*, 1977; 55: 88–94, as quoted in the *New England Journal of Medicine*, 1995; 333(4): 263.

29 B. Pomeranz and G. Stu, *Scientific Basis of Acupuncture* (New York: Springer-Verlag, 1989).

30 A. Colston Wentz, 'Infertility' (Book review), *New England Journal of Medicine*, 1995; 333(4): 263.
31 Becker and Selden, *The Body Electric*: 235.

CHAPTER FOUR: THE LANGUAGE OF THE CELL

1 J. Benveniste, B. Arnoux and L. Hadji, 'Highly dilute antigen increases coronary flow of isolated heart from immunized guinea-pigs', *FASEB Journal*, 1992; 6: A1610. Also presented at 'Experimental Biology – 98 (FASEB)', San Francisco, 20 April 1998.
2 M. Schiff, *The Memory of Water: Homeopathy and the Battle of New Ideas in the New Science* (HarperCollins, 1994): 22.
3 Ibid.: 26.
4 E. Davenas *et al.*, 'Human basophil degranulation triggered by very dilute antiserum against IgE', *Nature*, 1988; 333(6176): 816–8.
5 J. Maddox, 'Editorial', *Nature*, 1988; 333: 818; see also M. Schiff, *The Memory of Water*: 86.
6 J. Benveniste's reply to *Nature*, 1988; 334: 291. For a full account of the *Nature* visit, see J. Maddox, *et al.*, 'High-dilution experiments a delusion', *Nature*, 1988; 334: 287–90; J. Benveniste's reply to *Nature*; also Schiff, *Memory of Water*, chapter 6, pp. 85–95.
7 Schiff, *Memory of Water*: 57.
8 Ibid.: 103.
9 J. Benveniste, 'Understanding digital biology', unpublished position paper, June 14, 1998; also interviews with J. Benveniste, October 1999.
10 J. Benveniste, *et al.*, 'Digital recording/transmission of the cholinergic signal,' *FASEB Journal*, 1996, 10: A1479; Y. Thomas, *et al.*, 'Direct transmission to cells of a molecular signal (phorbol myristate acetate, PMA) via an electronic device,' *FASEB Journal*, 1995; 9: A227; J. Aïssa *et al.*, 'Molecular signalling at high dilution or by means of electronic circuitry', *Journal of Immunology*, 1993; 150: 146A; J. Aïssa, 'Electronic transmission of the cholinergic signal', *FASEB Journal*, 1995; 9: A683; Y. Thomas, 'Modulation of human neutrophil activation by "electronic" phorbol myristate acetate (PMA)', *FASEB Journal*, 1996; 10: A1479. (For a full listing of papers, see www.digibio.com).
11 J. Benveniste, P. Jurgens *et al.*, 'Transatlantic transfer of digitized antigen signal by telephone link', *Journal of Allergy and Clinical Immunology*, 1997; 99: S175.
12 Schiff, *Memory of Water*: 14–15.
13 D. Loye, *An Arrow Through Chaos: How We See into the Future* (Rochester, Vt: Park Street Press, 1983): 146.
14 J. Benveniste *et al.*, 'A simple and fast method for *in vivo* demonstration of electromagnetic molecular signaling (EMS) via high dilution or computer recording', *FASEB Journal*, 1999; 13: A163.
15 J. Benveniste *et al.*, 'The molecular signal is not functioning in the absence of "informed" water', *FASEB Journal*, 1999; 13: A163.
16 M. Jibu, S. Hagan, S. Hameroff *et al.*, 'Quantum optical coherence in cytoskeletal microtubules: implications for brain function', *BioSystems*, 1994; 32: 95–209.
17 A. H. Frey, 'Electromagnetic field interactions with biological systems', *FASEB Journal*, 1993; 7: 272.
18 M. Bastide *et al.*, 'Activity and chronopharmacology of very low doses of physiological immune inducers,' *Immunology Today*, 1985; 6: 234–5; L. Demangeat *et al.*, Mod-

ifications des temps de relaxation RMN à 4MHz des protons du solvant dans les très hautes dilutions salines de silice/lactose', *Journal of Medical Nuclear Biophysics*, 1992; 16: 135–45; B. J. Youbicier-Simo *et al.*, 'Effects of embryonic bursectomy and *in ovo* administration of highly diluted bursin on an adrenocorticotropic and immune response to chickens', *International Journal of Immunotherapy*, 1993; IX: 169–80; P. C. Endler *et al.*, 'The effect of highly diluted agitated thyroxine on the climbing activity of frogs', *Veterinary and Human Toxicology*, 1994; 36: 56–9.

19 P. C. Endler *et al.*, 'Transmission of hormone information by non-molecular means', *FASEB Journal*, 1994; 8: A400; F. Senekowitsch *et al.*, 'Hormone effects by CD record/replay', *FASEB Journal*, 1995; 9: A392.

20 *The Guardian*, March 15, 2001; see also J. Sainte-Laudy and P. Belon, 'Analysis of immunosuppressive activity of serial dilutions of histamines on human basophil activation by flow symmetry', *Inflammation Research*, 1996; Suppl 1: S33–4.

21 D. Reilly, 'Is evidence for homeopathy reproducible?' *The Lancet*, 1994; 344: 1601–6.

22 J. Jacobs, 'Homoeopathic treatment of acute childhood diarrhoea', *British Homoeopathic Journal*, 1993; 82: 83–6.

23 E. S. M. deLange deKlerk and J. Bloomer, 'Effect of homoeopathic medicine on daily burdens of symptoms in children with recurrent upper respiratory tract infections', *British Medical Journal*, 1994; 309: 1329–32.

24 F. J. Master, 'A study of homoeopathic drugs in essential hypertension', *British Homoeopathic Journal*, 1987; 76: 120–1.

25 D. Reilly, 'Is evidence for homeopathy reproducible?' *The Lancet*, 1994; 344: 1601-6.

26 Ibid.: 1585.

27 J. Benveniste, Letter, *The Lancet*, 1998; 351: 367.

28 Description of these results from a telephone conversation with Jacques Benveniste, November 10, 2000.

CHAPTER FIVE: RESONATING WITH THE WORLD

1 Description of Penrose's and Lashley's experiments from Karl Pribram, telephone interview, June 14, 2000;.M.Talbot, *The Holographic Universe* (New York: HarperCollins, 1991): 11-13.

2 K. Pribram, 'Autobiography in anecdote: the founding of experimental neuropsychology', in Robert Bilder, (ed.), *The History of Neuroscience in Autobiography* (San Diego, CA: Academic Press, 1998): 306–49.

3 Description of Lashley's laboratory protocol from Karl Pribram, telephone interview, June 14, 2000.

4 K. S. Lashley, *Brain Mechanisms and Intelligence* (Chicago: University of Chicago Press, 1929).

5 K. S. Lashley, 'In search of the engram', in Society for Experimental Biology, *Physiological Mechanisms in Animal Behavior* (New York: Academic Press, 1950): 501, as quoted in K. Pribram, *Languages of the Brain: Experimental Paradoxes and Principles in Neurobiology* (New York: Brandon House, 1971): 26.

6 Pribram, 'Autobiography'.

7 As quoted in K. Pribram, *Brain and Perception: Holonomy and Structure in Figural Processing* (Hillsdale, NJ: Lawrence Erlbaum, 1991): 9.

8 Talbot, *Holographic Universe*: 18–19.

9 D. Loye, *An Arrow Through Chaos* (Rochester, Vt: Park Street Press, 2000): 16–17.

10 Karl Pribram, telephone interview, June 14, 2000.
11 Various interviews with K. Pribram, June 2000; see also Talbot, *Holographic Universe*: 19.
12 Full description of his discovery as a result of an interview with Karl Pribram, London, September 9, 1999.
13 Pribram, 'Autobiography'.
14 Pribram, *Brain and Perception*: 27.
15 Pribram, *Brain and Perception*: Acknowledgments, xx; also, interview with Pribram, London, September 9, 1999.
16 Karl Pribram, telephone interviews, June 14 and July 7, 2000; also meeting in Liège, Belgium, August 12, 1999.
17 Loye, *Arrow Through Chaos*: 150.
18 Talbot, *Holographic Universe*: 21.
19 Correspondence with K. Pribram, July 5, 2001.
20 Talbot, *Holographic Universe*: 26.
21 R. DeValois and K. DeValois, *Spatial Vision* (Oxford: Oxford University Press, 1988).
22 Pribram, *Brain and Perception*: 76; also reviews of DeValois and DeValois, 'Spatial vision', *Annual Review of Psychology*, 1980: 309–41.
23 Pribram, *Brain and Perception*, chapter 9.
24 Pribram, *Brain and Perception*: 79.
25 Pribram, *Brain and Perception*: 76–7.
26 Pribram, *Brain and Perception*: 75.
27 Pribram, *Brain and Perception*: 137; see also Talbot, *Holographic Universe*: 27–30.
28 Ibid.
29 Telephone interviews with Karl Pribram, May 2000.
30 Pribram, *Brain and Perception*: 141.
31 W. J. Schempp, *Magnetic Resonance Imaging: Mathematical Foundations and Applications* (London: Wiley-Liss, 1998).
32 R. Penrose, *Shadows of the Mind: A Search for the Missing Science of Consciousness* (New York: Vintage, 1994): 367.
33 S. R. Hameroff, *Ultimate Computing: Biomolecular Consciousness and Nanotechnology* (Amsterdam: North Holland, 1987).
34 Ibid; also E. Laszlo, *The Interconnected Universe: Conceptual Foundations of Transdisciplinary Unified Theory* (Singapore: World Scientific, 1995): 41.
35 Pribram, *Brain and Perception*: 283.
36 M. Jibu and K. Yasue, 'A physical picture of Umezawa's quantum brain dynamics', in R. Trappl (ed.) *Cybernetics and Systems Research, '92* (Singapore: World Scientific, 1992); 'The basics of quantum brain dynamics', in K. H. Pribram (ed.) *Proceedings of the First Appalachian Conference on Behavioral Neurodynamics* (Radford: Center for Brain Research and Informational Sciences, Radford University, September 17–20, 1992); 'Intracellular quantum signal transfer in Umezawa's quantum brain dynamics', *Cybernetics Systems International*, 1993; 1(24): 1–7; 'Introduction to quantum brain dynamics', in E. Carvallo (ed.) *Nature, Cognition and System III* (London: Kluwer Academic, 1993).
37 C. D. Laughlin, 'Archetypes, neurognosis and the quantum sea', *Journal of Scientific Exploration*, 1996; 10: 375–400.
38 E. Insinna, correspondence and enclosures to author, November 5, 1998; also, E. Insinna 'Ciliated cell electrodynamics: from cilia and flagella to ciliated sensory sys-

tems', in A. Malhotra (ed.), *Advances in Structural Biology* (Stamford, Conn: JAI Press, 1999): 5.

39 M. Jibu, S. Hagan, S. Hameroff *et al.*, 'Quantum optical coherence in cytoskeletal microtubules: implications for brain function', *BioSystems*, 1994; 32: 95–209.

40 Ibid.

41 D. Zohar, *The Quantum Self* (London: Flamingo, 1991): 70.

42 Laszlo, *The Interconnected Universe*: 41.

43 Hameroff, *Ultimate computing*; Jibu *et al.*, 'Quantum optical coherence'.

44 E. Del Giudice *et al.*, 'Electromagnetic field and spontaneous symmetry breaking in biological matter', *Nuclear Physics*, 1983; B275(FS17): 185–99.

45 D. Bohm, *Wholeness and the Implicate Order* (London: Routledge, 1983).

46 Pribram has also postulated that humans also possess 'feedforward' loops of images and information which enable them to actively seek out specific information or stimuli: – looking for a mate of a certain type is just one example (correspondence with Karl Pribram, July 5, 2001. For full explanation, see also Dave Loye, *Arrow Through Chaos*: 22–3.

47 Laszlo, *Interconnected Universe*.

48 M. Jibu and K. Yasue, 'The basis of quantum brain dynamics', in K. H. Pribram (ed.), *Rethinking Neural Networks: Quantum Fields and Biological Data* (Hillsdale, NJ: Lawrence Erlbaum, 1993): 121–45

49 Laszlo, *Interconnected Universe*: 100–1.

50 Laughlin, 'Archetypes, neurognosis and the quantum sea'.

CHAPTER SIX: THE CREATIVE OBSERVER

1 For all history concerning Helmut Schmidt, correspondence with Helmut Schmidt, March 13, 1999; also telephone interviews with Schmidt, May 14, 2001, and May 16, 2001. See also R. S. Broughton, *Parapsychology: The Controversial Science* (New York: Ballantine, 1991).

2 Rhine eventually wrote his results in a book entitled *Extra-sensory Perception* (Boston: Bruce Humphries, 1964).

3 Telephone interview with Helmut Schmidt, May 16, 2001.

4 Interview with Robert Jahn and Brenda Dunne, Amsterdam, October 19, 2000; also R. G. Jahn and B. G. Dunne, *Margins of Reality: The Role of Consciousness in the Physical World* (New York: Harcourt, Brace, Jovanovich, 1987): 58–62.

5 E. Lazlo, *The Interconnected Universe: Conceptual Foundations of Transdisciplinary Unified Theory* (Singapore: World Scientific, 1995): 56.

6 H. Schmidt, 'Quantum processes predicted?', *New Scientist*, October 16, 1969: 114–15.

7 For amplification of this idea, see D. Radin and R. Nelson, 'Evidence for consciousness-related anomalies in random physical systems', *Foundations of Physics*, 1989; 19(12): 1499–514; also, D. Zohar, *The Quantum Self* (London: Flamingo, 1991): 33–4.

8 E. J. Squires, 'Many views of one world – an interpretation of quantum theory', *European Journal of Physics*, 1987; 8: 173.

9 H. Schmidt, 'Mental influence on random events', *New Scientist*, June 24, 1971; 757–8.

10 Broughton, *Parapsychology*: 177.

11 For the description of Helmut Schmidt's machine, correspondence with Schmidt, March 20, 1999; see also, Broughton, *Parapsychology*: 125–7; and D. Radin, *The Con-*

scious Universe: The Scientific Truth of Psychic Phenomena (New York: HarperEdge, 1997): 138–40.

12 Schmidt, 'Quantum processes'.

13 Schmidt, 'Mental influence'.

14 Ibid.

15 Telephone interview with Helmut Schmidt, May 14, 2001.

16 For the history of the PEAR program, interviews with Brenda Dunne, Princeton, June 23, 1998, and Robert Jahn and Brenda Dunne, Amsterdam, October 19, 2000.

17 Dunne and Jahn, *Margins of Reality*: 96–8.

18 R. G. Jahn *et al.*, 'Correlations of random binary sequences with prestated operator intention: a review of a 12-year program', *Journal of Scientific Exploration*, 1997; 11: 345–67.

19 Interview with Brenda Dunne, Amsterdam, October 19, 2000.

20 Jahn, 'Correlations': 350.

21 Ibid.

22 Radin and Nelson, 'Evidence for consciousness-related anomalies'; see also R. D. Nelson and D. I. Radin, 'When immovable objections meet irresistible evidence', *Behavioral and Brain Sciences*, 1987; 10: 600–1; 'Statistically robust anomalous effects: replication in random event generator experiments', in L. Henchle and R. E. Berger (eds), *RIP 1988* (Metuchen, NJ: Scarecrow Press, 1988): 23–6.

23 D. Radin and D. C. Ferrari, 'Effect of consciousness on the fall of dice: a meta-analysis', *Journal of Scientific Exploration*, 1991; 5: 61–84.

24 Broughton, *Parapsychology*: 177.

25 Radin, *Conscious Universe*: 140.

26 Radin and Nelson, 'Evidence for consciousness-related anomalies'.

27 D. Radin and R. Nelson, 'Meta-analysis of mind–matter interaction experiments, 1959–2000', unpublished, www.boundaryinstitute.org.

28 Radin and Nelson, 'Evidence for consciousness-related anomalies'.

29 R. D. Nelson, 'Effect size per hour: a natural unit for interpreting anomalous experiments', *PEAR Technical Note 94003*, September 1994.

30 W. Braud, 'Wellness implications of retroactive intentional influence: exploring an outrageous hypothesis', *Alternative Therapies*, 2000; 6(1): 37–48.

31 For the explanation and analogy of effect size, see Radin, *Conscious Universe*: 154–5; also W. Braud, 'Wellness implications'.

32 René Peoc'h, 'Psychokinetic action of young chicks on the path of an 'illuminated source', *Journal of Scientific Exploration*, 1995; 9(2): 223.

33 R. Jahn and B. Dunne, *Margins of Reality*: 242–59.

34 B. J. Dunne, 'Co-operator experiments with an REG device', *PEAR Technical Note 91005*, December 1991.

35 Interview with Brenda Dunne, Princeton, June 23, 1998.

36 Jahn and Dunne, *Margins*: 257.

37 Jahn *et al.*, Correlations: 356; also interview with Brenda Dunne, Princeton, June 23, 1998.

38 B. J. Dunne, 'Gender differences in human/machine anomalies', *Journal of Scientific Exploration*, 1998; 12(1): 3–55.

39 Interview with Brenda Dunne, Princeton, June 23, 1998.

40 Interview with Robert Jahn and Brenda Dunne, Amsterdam, October 19, 2000.

41 R. G. Jahn and B. J. Dunne, 'ArtREG: a random event experiment utilizing picture-

preference feedback', *Journal of Scientific Exploration*, 2000: 14(3): 383–409.

42 Interview with Robert Jahn and Brenda Dunne, Amsterdam, October 19, 2000.

43 R. Jahn, 'A modular model of mind/matter manifestations', *PEAR Technical Note 2001.01*, May 2001.

44 Ideas in this paragraph: discussion with Robert Jahn and Brenda Dunne, Amsterdam, October 19, 2000; also R. Jahn, 'Modular Model'.

45 Jahn and Dunne, 'Science of the subjective'.

CHAPTER SEVEN: SHARING DREAMS

1 Description of the Amazon indians from a study being conducted by The Institute of Noetic Sciences, which appeared in M. Schlitz, 'On consciousness, causation and evolution', *Alternative Therapies*, July 1998; 4(4): 82–90.

2 R. S. Broughton, *Parapsychology: The Controversial Science* (New York: Ballantine, 1991): 91–2.

3 Interview with William Braud, California, October 25, 1999.

4 Interview with William Braud, California, October 25, 1999.

5 D. Radin, *The Conscious Universe: The Scientific Truth of Psychic Phenomena* (HarperEdge: New York, 1997); also D. J. Bierman (ed.), *Proceedings of Presented Papers*, 37th Annual Parapsychological Association Convention, Amsterdam (Fairhaven, Mass.: Parapsychological Association, 1994): 71.

6 Broughton, *Parapsychology*: 98.

7 C. Tart, 'Physiological correlates of psi cognition', *International Journal of Parapsychology*, 1963: 5; 375–86; also interview with Charles Tart, California, October 29, 1999.

8 D. Delanoy, now of the University of Edinburgh, has carried out similar studies, e.g. D. Delanoy and S. Sah, 'Cognitive and psychological psi responses in remote positive and neutral emotional states', in Bierman (ed.), *Proceedings of Presented Papers*.

9 C. Tart, 'Psychedelic experiences associated with a novel hypnotic procedure: mutual hypnosis', in C. T. Tart (ed.), *Altered States of Consciousness* (New York: John Wiley, 1969): 291–308.

10 W. Braud and M. J. Schlitz, 'Consciousness interactions with remote biological systems: anomalous intentionality effects', *Subtle Energies*, 1991; 2(1): 1–46.

11 M. Schlitz and S. LaBerge, 'Autonomic detection of remote observation: two conceptual replications', in Bierman (ed.), *Proceedings of Presented Papers:* 465–78.

12 W. Braud *et al.*, 'Further studies of autonomic detection of remote staring: replication, new control procedures and personality correlates', *Journal of Parapsychology*, 1993; 57: 391–409. These studies were replicated by Schlitz and LaBerge, 'Autonomic detection'.

13 W. Braud and M. Schlitz, 'Psychokinetic influence on electrodermal activity', *Journal of Parapsychology*, 1983; 47(2): 95–119.

14 W. Braud *et al.*, 'Attention focusing facilitated through remote mental interaction', *Journal of the American Society for Psychical Research*, 1995; 89(2): 103–15.

15 M. Schlitz and W. Braud, 'Distant intentionality and healing: assessing the evidence', *Alternative Therapies*, 1997: 3(6): 62–73.

16 W. Braud and M. Schlitz, Psychokinetic influence on electrodermal activity', *Journal of Parapsychology*, 1983; 47: 95–119. Braud's studies were also independently replicated at the University of Edinburgh and the University of Nevada. D. Delanoy, 'Cognitive and physiological psi responses to remote positive and neutral emotional states', in Bierman (ed.), *Proceedings of Presented Papers*: 1298–38; also R. Wezelman

et al., 'An experimental test of magic: healing rituals', in E. C. May (ed.), *Proceedings of Presented Papers,* 39th Annual Parapsychological Association Convention, San Diego, Calif. (Fairhaven, Mass.: Parapsychological Association, 1996): 1–12.

17 W. Braud and M. Schlitz, 'A methodology for the objective study of transpersonal imagery', *Journal of Scientific Exploration*, 1989; 3(1): 43–63.

18 W. G. Braud, 'Psi-conducive states', *Journal of Communication*, 1975; 25(1): 142–52.

19 Broughton, *Parapsychology*: 103.

20 *Proceedings of the International Symposium on the Physiological and Biochemical Basis of Brain Activity*, St Petersburg, Russia, June 22–4, 1992; see also *Second Russian–Swedish Symposium on New Research in Neurobiology*, Moscow, Russia, May 19–21, 1992.

21 R. Rosenthal, 'Combining results of independent studies', *Psychological Bulletin*, 1978; 85: 185–93.

22 Radin, *Conscious Universe*: 79.

23 W. G. Braud, 'Honoring our natural experiences', *The Journal of the American Society for Psychical Research*, 1994; 88(3): 293–308.

24 Years later, this very idea became the subject of a book. L. Dossey's *Be Careful What you Pray For . . . You Just Might Get It* (HarperSanFrancisco, 1997) provides exhaustive examples of the power of negative thoughts to harm and also how to protect yourself from them.

25 W. G. Braud, 'Blocking/shielding psychic functioning through psychological and psychic techniques: a report of three preliminary studies', in R. White and I. Solfvin (eds), *Research in Parapsychology*, 1984 (Metuchen, NY: Scarecrow Press, 1985): 42–4.

26 W. G. Braud, 'Implications and applications of laboratory psi findings', *European Journal of Parapsychology*, 1990–91; 8: 57–65.

27 W. Braud *et al.*, 'Further studies of the bio-PK effect: feedback, blocking, generality/specificity', in White and Solfvin (eds), *Research in Parapsychology*: 45–8.

28 D. Bohm, *Wholeness and the Implicate Order* (London: Routledge, 1980).

29 E. Laszlo, *The Interconnected Universe: Conceptual Foundations of Transdisciplinary Unified Theory* (Singapore: World Scientific, 1995): 101.

30 J. Grinberg-Zylberbaum and J. Ramos, 'Patterns of interhemisphere correlations during human communication', *International Journal of Neuroscience*, 1987; 36: 41–53; J. Grinberg-Zylberbaum *et al.*, 'Human communication and the electrophysiological activity of the brain', *Subtle Energies*, 1992; 3(3): 25–43.

31 These have been explored in detail by Ian Stevenson; see I. Stevenson, *Children Who Remember Previous Lives* (Charlottesville, Va: University Press of Virginia, 1987).

32 Laszlo, *Interconnected Universe*: 102–3.

33 Braud, *Honoring Our Natural Experiences*.

34 Indeed, Marilyn Schlitz and Charles Honorton carried out an experiment showing that artistically gifted individuals were better at ESP than the ordinary population. See M. J. Schlitz and C. Honorton, 'Ganzfeld psi performance within an artistically gifted population', *The Journal of the American Society for Psychical Research*, 1992; 86(2): 83–98.

35 L. F. Berkman and S. L. Syme, 'Social networks, host resistance and mortality: a nine-year follow-up study of Alameda County residents,' *American Journal of Epidemiology*, 1979; 109(2): 186–204.

36 L. Galland, *The Four Pillars of Healing* (New York: Random House, 1997): 103–5.

CHAPTER EIGHT: THE EXTENDED EYE

1 C. Backster, 'Evidence of a primary perception in plant life', *International Journal of Parapsychology*, 1967; X: 141. Hal's paper 'Toward a quantum theory of life process', written in 1972, was never published. 'With 30 years' hindsight, and the lack of unambiguous verification of either the Backster effect or tachyons – the two lynchpins of this proposal – it seems somewhat naïve. But it got me started,' wrote Puthoff to the author on March 15, 2000. He also notes: 'By the way, I never did get to do the proposed experiment.'

2 H. Puthoff, 'Toward a quantum theory of life process'.

3 G. R. Schmeidler, 'PK effects upon continuously recorded temperatures', *Journal of the American Society of Psychical Research*, 1997; 67(4), cited in H. Puthoff and R. Targ, 'A perceptual channel for information transfer over kilometer distances: historical perspective and recent research', *Proceedings of the IEEE*, 1976; 64(3): 329–54.

4 S. Ostrander and L. Schroeder, *Psychic Discoveries Behind the Iron Curtain* (now abridged in *Psychic Discoveries*, New York: Marlowe & Company, 1997), published in 1971, caused a flood of concern about so-called 'psychic warfare'.

5 J. Schnabel, *Remote Viewers: The Secret History of America's Psychic Spies* (New York: Dell, 1997): 94–5.

6 Hank Turner is a pseudonym of a CIA employee referred to as 'Bill O'Donnell' in Schnabel's book.

7 For entire description of the West Virginia military installation facility and Pat Price, see Schnabel, *Remote Viewers*: 104–13.

8 H. Puthoff and R. Targ, 'Final report, covering the period January 1974–February 1975 Part II–Research Report, December 1, 1975, *Perceptual Augmentation Techniques*, SRI Project 3183; also H. E. Puthoff, 'CIA-initiated remote viewing program at Stanford Research Institute, *Journal of Scientific Exploration*, 1996; 10(1): 63–75.

9 R. Targ, *Miracles of Mind: Exploring Nonlocal Consciousness and Spiritual Healing* (Novato, Calif: New World Library, 1999): 46–7; D. Radin, *The Conscious Universe: The Scientific Truth of Psychic Phenomena* (New York: HarperEdge, 1997): 25–6.

10 C. A. Robinson, Jr, 'Soviets push for beam weapon', *Aviation Week*, May 2, 1977.

11 Interview with Edwin May, California, October 25, 1999.

12 H. Puthoff, 'CIA-initiated remote viewing program at Stanford Research Institute'.

13 Interview with Hal Puthoff, January 20, 2000; also Schnabel, *Remote Viewers*.

14 H. Puthoff, 'Experimental psi research: implication for physics', in R. Jahn (ed.), *The Role of Consciousness in the Physical World*, AAA Selected Symposia Series (Boulder, Colorado: Westview Press, 1981): 41.

15 R. Targ and H. Puthoff, *Mind-Reach: Scientists Look at Psychic Ability* (New York: Delacorte Press, 1977): 50.

16 Schnabel, *Remote Viewers*: 142.

17 Puthoff and Targ, 'Perceptual channel': 342.

18 Ibid.: 338.

19 Ibid.: 330–1.

20 Ibid.: 336.

21 B. Dunne and J. Bisaha, 'Precognitive remove viewing in the Chicago area: a replication of the Stanford experiment', *Journal of Parapsychology*, 1979; 43:17–30.

22 Radin, *Conscious Universe*: 105.

23 L. M. Kogan, 'Is telepathy possible?' *Radio Engineering*, 1966; 21 (Jan): 75, quoted in Puthoff and Targ, 'Perceptual channel': 329–53.
24 H. Puthoff and R. Targ, 'Final report, covering the period January 1974–February 1975 Part II–Research Report, December 1, 1975, *Perceptual Augmentation Techniques*, SRI Project 3183: 58.
25 Telephone interview with Hal Puthoff, January 20, 2000; see also Targ and Puthoff, *Mind-Reach*.
26 Schnabel, *Remote Viewers*: 74–5.
27 Interview with Edwin May and Dean Radin, California, October 25, 1999.
28 Various telephone interviews with Hal Puthoff, August 2000.
29 J. Utts, 'An assessment of the evidence for psychic functioning', *Journal of Scientific Exploration*, 1996; 10: 3–30.

CHAPTER NINE: THE ENDLESS HERE AND NOW

1 R. Targ and J. Katra, *Miracles of Mind: Exploring Nonlocal Consciousness and Spiritual Healing* (Novato, Calif: New World Library, 1999): 42–4.
2 B. J. Dunne and R. G. Jahn, 'Experiments in remote human/machine interaction', *Journal of Scientific Exploration*, 1992; 6(4): 311–32.
3 In all the SRI experiments, they never found a limit to the distance over which the channel worked. Many years later, in an ironical reversal of the SRI studies, Russell Targ would have a Russian psychic in Moscow do a remote viewing of an unknown target site in San Francisco. Djuna Davitashvili, a noted Russian psychic healer, who had never done remote-viewing experiments before, was asked to describe where a colleague of theirs was at the time in a location in San Francisco even unknown to Targ. After being shown his photo, she correctly described a plaza with a merry-go-round (eventually Targ was told that the colleague was standing in front of one at a plaza on San Francisco's Pier 39). The picture she drew of both the plaza and of the carousel's horses bore a remarkable similarity to the actual site. For a full account, see R. Targ and J. Katra, *Miracles of Mind*: 29–36.
4 For the Chicago, Arizona and Moscow remote-viewing experiment, R. G. Jahn and B. J. Dunne, *Margins of Reality* (New York: Harcourt Brace Jovanovich, 1987): 162– 7.
5 For the NASA and irrigation-ditch examples, Jahn and Dunne, *Margins*: 188.
6 D. Radin, *The Conscious Universe: The Scientific Truth of Psychic Phenomena* (New York: HarperEdge, 1997): 113–4; R. Broughton, *Parapsychology: The Controversial Science* (New York: Ballantine, 1991): 292.
7 For an excellent summary of this and other precognitive studies, see Radin, *The Conscious Universe*: 111–25.
8 R. S. Broughton, *Parapsychology*: 95–7.
9 Ibid.: 98. Maimonides wasn't the first to scientifically document dreams. In the early part of this century, J. W. Dunne conducted experiments with subjects and their dreams, scientifically demonstrating that what people dreamed largely came true. J. W. Dunne, *An Experiment in Time* (London: Faber, 1926).
10 As it happened, Radin's expectation that he'd reached a safe haven to carry out his research was premature. As soon as he published a book on psychic research and began to attract some media attention, the University refused to renew his contract. He was left to find work in privately funded research projects. At the time of writing, he is working at the Institute of Noetic Sciences.

11 For a full description of the Radin experiment, see Radin, *Conscious Universe*: 119–24.
12 D. J. Bierman and D. I. Radin, 'Anomalous anticipatory response on randomized future conditions', *Perceptual and Motor Skills*, 1997; 84: 689–90.
13 D. J. Bierman, 'Anomalous aspects of intuition', paper presented at the Fourth Biennial European meeting of the Society for Scientific Exploration, Valencia, October 9-11, 1998; also interview with Professor Bierman, Valencia, October 9, 1998.
14 D. I. Radin and E. C. May, 'Testing the intuitive data sorting model with pseudorandom number generators: a proposed method', in D. H. Weiner and R. G. Nelson (eds), *Research in Parapsychology 1986* (Metuchen, NJ: Scarecrow, 1987): 109–11. For a description of the test, see Broughton, *Parapsychology*: 137–9.
15 Broughton, *Parapsychology*: 175–6; also telephone interviews with Helmut Schmidt, May 2001.
16 H. Schmidt, 'Additional affect for PK on pre-recorded targets', *Journal of Parapsychology*, 1985; 49: 229–44; 'PK tests with and without preobservation by animals', in L. S. Henkel and J. Palmer (eds), *Research in Parapsychology 1989* (Metuchen, NJ: Scarecrow Press, 1990): 15–9, in W. Braud, 'Wellness implications of retroactive intentional influence: exploring an outrageous hypothesis', *Alternative Therapies*, 2000, 6(1): 37–48.
17 R. G. Jahn *et al*., 'Correlations of random binary sequences with pre-stated operator intention: a review of a 12-year program', *Journal of Scientific Exploration*, 1997; 11(3): 345–67.
18 Braud, 'Wellness implications'.
19 J. Gribbin, *Q Is for Quantum: Particle Physics from A to Z* (Phoenix, 1999): 531–4.
20 Radin, various telephone interviews in 2001.
21 E. Laszlo, *The Interconnected Universe, Conceptual Foundations of Transdisciplinary Unified Theory* (Singapore: World Scientific, 1995): 31.
22 D. Bohm, *Wholeness and the Implicate Order* (London: Routledge, 1980): 211.
23 Ibid.
24 Braud, 'Wellness implications'.

CHAPTER TEN: THE HEALING FIELD

1 Interview with Elisabeth Targ, California, October 28, 1999.
2 Ibid.
3 Both experiments, B. Grad, 'Some biological effects of "laying-on of hands": a review of experiments with animals and plants', *Journal of the American Society for Psychical Research*, 1965; 59: 95–127.
4 L. Dossey, *Be Careful What You Pray For . . . You Just Might Get It* (HarperSanFrancisco, 1997): 179.
5 B. Grad, 'Dimensions in "Some biological effects of the laying on of hands" and their implications,' in H. A. Otto and J. W. Knight (eds), *Dimensions in Wholistic Healing: New Frontiers in the Treatment of the Whole Person* (Chicago: Nelson-Hall, 1979): 199–212.
6 B. Grad, R. J. Cadoret and G. K. Paul, 'The influence of an unorthodox method of treatment on wound healing in mice', *International Journal of Parapsychology*, 1963; 3: 5–24.
7 B. Grad, 'Healing by the laying on of hands: review of experiments and implications', *Pastoral Psychology*, 1970; 21: 19–26.

8 F. W. J. Snel and P. R. Hol, 'Psychokinesis experiments in casein induced amyloidosis of the hamster', *Journal of Parapsychology*, 1983; 5(1): 51–76; Grad, 'Some biological effects of laying on of hands'; F. W. J. Snel and P. C. Van der Sijde, 'The effect of paranormal healing on tumor growth', *Journal of Scientific Exploration*, 1995; 9(2): 209–21. See also E. Targ, 'Evaluating distant healing: a research review, *Alternative therapies*, 1997; 3:748.

9 J. Barry, 'General and comparative study of the psychokinetic effect on a fungus culture', *Journal of Parapsychology*, 1968; 32: 237–43; E. Haraldsson and T. Thorsteinsson, 'Psychokinetic effects on yeast: an exploratory experiment', in W. G. Roll, R. L. Morris and J. D. Morris (eds), *Research in Parapsychology* (Metuchen, NJ: Scarecrow Press, 1972): 20–1; F. W. J. Snel, 'Influence on malignant cell growth research', *Letters of the University of Utrecht*, 1980; 10: 19–27.

10 C. B. Nash, 'Psychokinetic control of bacterial growth', *Journal of the American Society for Psychical Research*, 1982; 51: 217–21.

11 G. F. Solfvin, 'Psi expectancy effects in psychic healing studies with malarial mice', *European Journal of Parapsychology*, 1982; 4(2): 160–97.

12 R. Stanford, '"Associative activation of the unconscious" and "visualization" as methods for influencing the PK target', *Journal of the American Society for Psychical Research*, 1969; 63: 338–51.

13 R. N. Miller, 'Study on the effectiveness of remote mental healing', *Medical Hypotheses*, 1982; 8: 481–90.

14 R. C. Byrd, 'Positive therapeutic effects of intercessory prayer in a coronary care unit population', *Southern Medical Journal*, 1988; 81(7): 826–9.

15 B. Greyson, 'Distance healing of patients with major depression', *Journal of Scientific Exploration*, 1996; 10(4): 447–65.

16 F. Sicher and E. Targ *et al.*, 'A randomized double-blind study of the effect of distant healing in a population with advanced AIDS: report of a small scale study', *Western Journal of Medicine*, 1998; 168(6): 356–63.

17 W. Harris *et al.*, 'A randomized, controlled trial of the effects of remote, intercessory prayer on outcomes in patients admitted to the coronary care unit', *Archives of Internal Medicine*, 1999; 159 (19): 2273–8.

18 Interviews with E. Targ in California and on the telephone, October 28, 1999, and March 6, 2001.

19 Harris *et al.*, 'A randomized, controlled trial of the effects of remote, intercessory prayer'.

20 J. Barrett, 'Going the distance', *Intuition*, 1999; June/July: 30–1.

21 E. E. Green, 'Copper Wall research psychology and psychophysics: subtle energies and energy medicine: emerging theory and practice', *Proceedings*, First Annual Conference, International Society for the Study of Subtle Energies and Energy Medicine (ISSSEEM), Boulder, Colorado, June 21–25, 1991.

22 Summaries of studies of Qigong healing energy and information about the Qigong Database, a computerized resource center of published research on Qigong healing, in L. Dossey, *Be Careful What You Pray For*: 175–7.

23 R. D. Nelson, 'The physical basis of intentional healing systems', *PEAR Technical Note, 99001*, January 1999.

24 G. A. Kaplan, *et al.*, 'Social connections and morality from all causes and from cardiovascular disease: perspective evidence from Eastern Finland', *American*

Journal of Epidemiology, 1988; 128: 370–80.
25 D. Reed, *et al.*, 'Social networks and coronary heart disease among Japanese men in Hawaii', *American Journal of Epidemiology*, 1983; 117: 384–96; M. A. Pascucci and G. L. Loving, 'Ingredients of an old and healthy life: centenarian perspective', *Journal of Holistic Nursing*, 1997; 15: 199–213.
26 G. Schwarz, *et al.*, 'Accuracy and replicability of anomalous after-death communication across highly skilled mediums', *Journal of the Society for Psychical Research*, 2001; 65: 1–25.

CHAPTER ELEVEN: TELEGRAM FROM GAIA

1 For all material about the O.J. Simpson trial: *London Sunday Times* archives. Trial transcripts for the verdict day: the Associated Press's statistics of the O.J. Simpson trial.
2 Interview with Brenda Dunne at Princeton, June 28, 1998.
3 R. D. Nelson *et al.*, 'FieldREG anomalies in group situations', *Journal of Scientific Exploration*, 1996; 10(1): 111–41.
4 Ibid.
5 Ibid.
6 Ibid; also correspondence with R. Nelson, July 26, 2001.
7 R. D. Nelson and E. L. Mayer, 'A FieldREG application at the San Francisco Bay Revels, 1996', as reported in D. Radin, *The Conscious Universe: The Scientific Truth of Psychic Phenomena* (New York: HarperEdge, 1997): 171.
8 Nelson, 'FieldREG anomalies': 136.
9 R. D. Nelson *et al.*, 'FieldREGII: consciousness field effects: replications and explorations', *Journal of Scientific Exploration*, 1998; 12(3): 425–54.
10 For the entire study in Egypt: R. Nelson, 'FieldREG measurements in Egypt: resonant consciousness at sacred sites', Princeton Engineering Anomalies Research, School of Engineering/ Applied Science, *PEAR Technical Note* 97002, July 1997; telephone interview with Roger Nelson, February 2, 2001; also Nelson *et al.*, 'Field-REGII'.
11 For all descriptions of Dean Radin's experiments in this chapter, I am indebted to his excellent account of his own work in *The Conscious Universe*: 157–74. See also D. I. Radin, J. M. Rebman and M. P. Cross, 'Anomalous organization of random events by group consciousness: two exploratory experiments', *Journal of Scientific Exploration*, 1996; 10: 143–68.
12 D. Vaitl, 'Anomalous effects during Richard Wagner's operas', paper presented at the Fourth Biennial European Meeting of the Society for Scientific Exploration, Valencia, Spain, October 9–11, 1998.
13 Ibid.
14 D. Bierman, 'Exploring correlations between local emotional and global emotional events and the behavior of a random number generator', *Journal of Scientific Exploration*, 1996; 10: 363–74.
15 R. Nelson, 'Wishing for good weather: a natural experiment in group consciousness', *Journal of Scientific Exploration*, 1997; 11(1): 47–58.
16 J. S. Hagel, *et al.*, 'Effects of group practice of the Transcendental Meditation Program on preventing violent crime in Washington DC: results of the National Demonstration Project, June–July, 1993,' *Social Indicators Research*, 1994; 47: 153–201.
17 M. C. Dillbeck *et al.*, 'The Transcendental Meditation program and crime rate change in a sample of 48 cities', *Journal of Crime and Justice*, 1981; 4: 25–45.

18 D. W. Orme-Johnson *et al.*, 'International peace project in the Middle East: the effects of the Maharishi technology of the unified field', *Journal of Conflict Resolution*, 1988; 32: 776–812.
19 J. Lovelock, *Gaia: a New Look at Life on Earth* (Oxford: Oxford University Press, 1979).
20 R. Nelson *et al.*, 'Global resonance of consciousness: Princess Diana and Mother Teresa', *Electronic Journal of Parapsychology*, 1998.
21 Telephone interview with R. Nelson, February 2, 2001.
22 'Terrorist Disaster, September 11, 2001,' Global Consciousness Project website: http://noosphere.princeton.edu.
23 N. A. Klebanoff and P. K. Keyser, 'Menstrual synchronization: a qualitative study', *Journal of Holistic Nursing*, 1996; 14(2): 98–114.
24 In a speech in 1999 in Liège, Belgium, Mitchell would cite a little known report recording the experiences of Russian cosmonauts living aboard the Mir spacecraft for six months. Like Mitchell, they also experienced extraordinary perceptions in their waking and dream states, including precognition. It may well be that a long-duration space voyage provides some extraordinary means of tapping into The Field. S. V. Krichevskii, 'Extraordinary fantastic states/dreams of the astronauts in near-earth orbit: a new cosmic phenomenon', *Sozn Fiz Real*, 1996; 1(4): 60-9.

CHAPTER TWELVE: THE ZERO POINT AGE

1 Interview with Richard Obousy, Brighton, January 20, 2001.
2 Confirmed by Graham Ennis at Propulsion Workshop, Brighton, January 20, 2001.
3 C. Sagan, *Contact* (London: Orbit, 1997).
4 R. Forward, 'Extracting electrical energy from the vacuum by cohesion of charged foliated conductors', *Physical Review B*, 1984: 30: 1700.
5 H. Puthoff, 'Space propulsion: can empty space itself provide a solution?' *Ad Astra*, 1997; 9(1): 42–6.
6 R. Matthews, 'Nothing like a vacuum', *New Scientist,* February 25, 1995: 33.
7 Ibid.
8 H. Puthoff, quoted in *The Observer*, January 7, 2001: 13.
9 Telephone and in-person interviews with Hal Puthoff, January 2001.
10 Hal Puthoff, 'SETI: the velocity of light limitation and the Alcubierre warp drive: an integrating overview', *Physics Essays*, 1996; 9(1): 156–8.
11 H. Puthoff, 'Everything for nothing', *New Scientist*, July 28, 1990: 52–5.
12 H. Puthoff, interview, Brighton, January 20, 2001.
13 Quoted on the Propulsion Workshop website: www.workshop.cwc.net.
14 J. Benveniste, 'Specific remote detection for bacteria using an electromagnetic/digital procedure', *FASEB Journal*, 1999; 13: A852.
15 E. Mitchell, 'Nature's mind', keynote address, CASYS 1999, Liège, Belgium, August 8, 2000.
16 H. Puthoff, 'Far out ideas grounded in real physics', *Jane's Defence Weekly*, July 26, 2000; 34(4): 42–6.
17 Ibid.
18 P. W. Milonni, 'Semi-classical and quantum electrodynamical approaches in nonrelativistic radiation theory', *Physics Reports*, 1976; 25: 1–8.

BIBLIOGRAPHY

Abraham, R., McKenna, T. and Sheldrake, R., *Trialogues at the Edge of the West: Chaos, Creativity and the Resacralization of the World* (Santa Fe, NM: Bear, 1992).

Adler, R. *et al.*, 'Psychoneuroimmunology: interactions between the nervous system and the immune system', *Lancet*, 1995; 345: 99-103.

Adler, S. (in a selection of short articles dedicated to the work of Andrei Sakharov), 'A key to understanding gravity', *New Scientist*, April 30, 1981: 277-8.

Aïssa, J. *et al.*, 'Molecular signalling at high dilution or by means of electronic circuitry', *Journal of Immunology*, 1993; 150: 146A.

Aissa, J., 'Electronic transmission of the cholinergic signal', *FASEB Journal*, 1995; 9: A683.

Arnold, A., *The Corrupted Sciences* (London: Paladin, 1992).

Atmanspacher, H., 'Deviations from physical randomness due to human agent intention?', *Chaos, Solitons and Fractals*, 1999; 10(6): 935-52.

Auerbach, L., *Mind Over Matter: A Comprehensive Guide to Discovering Your Psychic Powers* (New York: Kensington, 1996).

Backster, C., 'Evidence of a primary perception in plant life', *International Journal of Parapsychology*, 1967; X: 141.

Ballentine, R., *Radical Healing: Mind-Body Medicine at its Most Practical and Transformative* (London: Rider, 1999).

Bancroft, A., *Modern Mystics and Sages* (London: Granada, 1978).

Barrett, J., 'Going the distance', *Intuition*, 1999; June/July: 30-1.

Barrow, J. D., *Impossibility: The Limits of Science and the Science of Limits* (Oxford: Oxford University Press, 1998).

Barrow, J., *The Book of Nothing* (London: Jonathan Cape, 2000).

Barry, J., 'General and comparative study of the psychokinetic effect on a fungus culture', *Journal of Parapsychology*, 1968; 32: 237-43.

Bastide, M., *et al.*, 'Activity and chronopharmacology of very low doses of physiological immune inducers', *Immunology Today*, 1985; 6: 234-5.

Becker, R. O., *Cross Currents: The Perils of Electropollution, the Promise of Electromedicine* (New York: Jeremy F. Tarcher/Putnam, 1990).

Becker, R. O. and Selden, G., *The Body Electric: Electromagnetism and the Foundation of Life* (London: Quill/William Morrow, 1985).

Behe, M. J., *Darwin's Black Box: The Biochemical Challenge to Evolution* (New York: Touchstone, 1996).

Benor, D. J., 'Survey of spiritual healing research', *Complementary Medical Research*, 1990; 4: 9-31.

Benor, D. J., *Healing Research*, vol.4 (Deddington, Oxfordshire: Helix Editions, 1992).

Benstead, D. and Constantine, S., *The Inward Revolution* (London: Warner, 1998).

Benveniste, J., 'Reply', *Nature*, 1988; 334: 291.

Benveniste, J., 'Reply (to Klaus Linde and coworkers) "Homeopathy trials going nowhere"', *Lancet*, 1997; 350: 824', Lancet, 1998; 351: 367.

Benveniste, J., 'Understanding digital biology', unpublished position paper, June 14, 1998.

Benveniste, J., 'From water memory to digital biology', *Network: The Scientific and Medical Network Review*, 1999; 69: 11-14.

Benveniste, J., 'Specific remote detection for bacteria using an electromagnetic/digital procedure', *FASEB Journal*, 1999; 13: A852.

Benveniste, J., Arnoux, B. and Hadji, L., 'Highly dilute antigen increases coronary flow of isolated heart from immunized guinea-pigs', *FASEB Journal*, 1992; 6: A1610. Also presented at 'Experimental Biology- 98 (FASEB)', San Francisco, April 20, 1998.

Benveniste, J., Jurgens, P. *et al.*, 'Transatlantic transfer of digitized antigen signal by telephone link', *Journal of Allergy and Clinical Immunology*, 1997; 99: S175.

Benveniste, J. *et al.*, 'Digital recording/transmission of the cholinergic signal', *FASEB Journal*, 1996; 10: A1479.

Benveniste, J. *et al.*, 'Digital biology: specificity of the digitized molecular signal', *FASEB Journal*, 1998; 12: A412.

Benveniste, J. *et al.*, 'A simple and fast method for *in vivo* demonstration of electromagnetic molecular signaling (EMS) via high dilution or computer recording', *FASEB Journal*, 1999; 13: A163.

Benveniste, J. *et al.*, 'The molecular signal is not functioning in the absence of "informed" water', *FASEB Journal*, 1999; 13: A163.

Berkman, L. F. and Syme, S. L., 'Social networks, host resistance and mortality: a nine-year follow-up study of Alameda County residents', *American Journal of Epidemiology*, 1979; 109(2): 186-204.

Bierman, D. J. (ed.), *Proceedings of Presented Papers*, 37th Annual Parapsychological Association Convention, Amsterdam (Fairhaven, Mass.: Parapsychological Association, 1994).

Bierman, D., 'Exploring correlations between local emotional and global emotional events and the behavior of a random number generator', *Journal of Scientific Exploration*, 1996; 10: 363-74.

Bierman, D. J., 'Anomalous aspects of intuition', paper presented at the Fourth Biennial European Meeting of the Society for Scientific Exploration, Valencia, Spain, October 9-11, 1998.

Bierman, D. J. and Radin, D. I., 'Anomalous anticipatory response on randomized future conditions', *Perceptual and Motor Skills*, 1997; 84: 689-90.

Bischof, M., 'The fate and future of field concepts - from metaphysical origins to holistic understanding in the biosciences', lecture given at the Fourth Biennial European Meeting of the Society for Scientific Exploration, Valencia, Spain, October 9-11, 1998.

Bischof, M., 'Holism and field theories in biology: non-molecular approaches and their relevance to biophysics', in J. J. Clang *et al.* (eds), *Biophotons* (Amsterdam: Kluwer Academic, 1998): 375-94.

Blom-Dahl, C. A., 'Precognitive remote perception and the third source paradigm', paper presented at the Fourth Biennial European Meeting of the Society for Scientific Exploration, Valencia, Spain, October 9-11, 1998.

Bloom, W. (ed.), *The Penguin Book of New Age and Holistic Writing* (Harmondsworth: Penguin, 2000).

Bohm, D., *Wholeness and the Implicate Order* (London: Routledge, 1980).

Boyer, T., 'Deviation of the blackbody radiation spectrum without quantum physics', *Physical Review*, 1969; 182: 1374.

Braud, W. G., 'Psi-conducive states', *Journal of Communication*, 1975; 25(1): 142-52.

Braud, W. G., 'Psi conducive conditions: explorations and interpretations', in B. Shapin and L. Coly (eds), *Psi and States of Awareness*, Proceedings of an International Conference held in Paris, France, August 24-26, 1977.

Braud, W. G., 'Blocking/shielding psychic functioning through psychological and psychic techniques: a report of three preliminary studies', in R. White and I. Solfvin (eds), *Research in Parapsychology*, 1984 (Metuchen, NJ: Scarecrow Press, 1985): 42-4.

Braud, W. G., 'On the use of living target systems in distant mental influence research', in L. Coly and J. D. S. McMahon (eds), *Psi Research Methodology: A Re-Examination*, Proceedings of an international conference held in Chapel Hill, North Carolina, October 29-30, 1988.

Braud, W. G., 'Distant mental influence of rate of hemolysis of human red blood cells', *Journal of the American Society for Psychical Research*, 1990; 84(1): 1-24.

Braud, W. G., 'Implications and applications of laboratory psi findings', *European Journal of Parapsychology*, 1990-91; 8: 57-65.

Braud, W. G., 'Reactions to an unseen gaze (remote attention): a review, with new data on autonomic staring detection', *Journal of Parapsychology* 1993; 57: 373-90.

Braud, W. G., 'Honoring our natural experiences', *Journal of the American Society for Psychical Research*, 1994; 88(3): 293-308.

Braud, W. G., 'Reaching for consciousness: expansions and complements', *Journal of the American Society for Psychical Research*, 1994; 88(3): 186-206.

Braud, W. G., 'Wellness implications of retroactive intentional influence: exploring an outrageous hypothesis', *Alternative Therapies*, 2000; 6(1): 37-48.

Braud, W. G. and Schlitz, M., 'Psychokinetic influence on electrodermal activity', *Journal of Parapsychology* 1983; 47(2): 95-119.

Braud, W. G. and Schlitz, M., 'A methodology for the objective study of transpersonal imagery', *Journal of Scientific Exploration*, 1989; 3(1): 43-63.

Braud, W. G. and Schlitz, M., 'Consciousness interactions with remote biological systems: anomalous intentionality effects', *Subtle Energies*, 1991; 2(1): 1-46.

Braud, W. *et al.*, 'Further studies of autonomic detection of remote staring: replication, new control procedures and personality correlates', *Journal of Parapsychology*, 1993; 57: 391-409.

Braud, W. *et al.*, 'Attention focusing facilitated through remote mental interaction', *Journal of the American Society for Psychical Research*, 1995; 89(2): 103-15.

Braud, W. *et al.*, 'Further studies of the bio-PK effect: feedback, blocking, generality/specificity', in R. White and J. Solfvin (eds), *Research in Parapsychology* 1984 (Metuchen, NJ: Scarecrow Press, 1985): 45-8.

Brennan, B. A., *Hands of Light: A Guide to Healing Through the Human Energy Field* (New York: Bantam, 1988).
Brennan, J. H., *Time Travel: A New Perspective* (St. Paul, Minn.: Llewellyn, 1997).
Broughton, R. S., *Parapsychology: The Controversial Science* (New York: Ballantine, 1991).
Brown, G., *The Energy of Life: The Science of What Makes our Minds and Bodies Work* (New York: Free Press/Simon & Schuster, 1999).
Brockman, J., *The Third Culture: Beyond the Scientific Revolution* (New York: Simon & Schuster, 1995).
Buderi, R., *The Invention that Changed the World: The Story of Radar from War to Peace* (London: Abacus, 1998).
Bunnell, T., 'The effect of hands-on healing on enzyme activity', *Research in Complementary Medicine*, 1996; 3: 265-40: 314; 3rd Annual Symposium on Complementary Health Care, Exeter, December 11-13, 1996.
Burr, H., *The Fields of Life* (New York: Ballantine, 1972).
Byrd, R. C., 'Positive therapeutic effects of intercessory prayer in a coronary care unit population', *Southern Medical Journal*, 1988; 81(7): 826-9.
Capra, F., *The Turning Point: Science, Society and the Rising Culture* (London: Flamingo, 1983).
Capra, F., *The Tao of Physics: An Explanation of the Parallels Between Modern Physics and Eastern Mysticism* (London: Flamingo, 1991).
Capra, F., *The Web of Life: A New Synthesis of Mind and Matter* (London: Flamingo, 1997).
Carey, J., *The Faber Book of Science* (London: Faber & Faber, 1995).
Chaikin, A., *A Man on the Moon: The Voyages of the Apollo Astronauts* (Harmondsworth: Penguin, 1998).
Chopra, D., *Quantum Healing: Exploring the Frontiers of Mind/Body Medicine* (New York: Bantam, 1989).
Clarke, A. C., 'When will the real space age begin?', *Ad Astra*, May/June 1996:13-15.
Clarke, A. C., *3001: The Final Odyssey* (London: HarperCollins, 1997).
Coats, C., *Living Energies: An Exposition of Concepts Related to the Theories of Victor Schauberger* (Bath: Gateway, 1996).
Coen, E., *The Art of Genes: How Organisms Make Themselves* (Oxford: Oxford University Press, 1999).
Cohen, S. and Popp, F. A., 'Biophoton emission of the human body', *Journal of Photochemistry and Photobiology B: Biology* 1997; 40:187-9.
Coghill, R. W., *Something in the Air* (Coghill Research Laboratories, 1998).
Coghill, R. W., *Electrohealing: The Medicine of the Future* (London: Thorsons, 1992).
Cole, D. C. and Puthoff, H. E., 'Extracting energy and heat from the vacuum', *Physical Review E*, 1993; 48(2): 1562-65.
Cornwell, J., *Consciousness and Human Identity* (Oxford: Oxford University Press, 1998).
Damasio, A. R., *Descartes' Error: Emotion, Reason and the Human Brain* (New York: G. P. Putnam, 1994).
Davelos, J., *The Science of Star Wars* (New York: St Martin's Press, 1999).
Davenas, E. *et al.*, 'Human basophil degranulation triggered by very dilute antiserum against IgE', *Nature*, 1988; 333(6176): 816-18.
Davidson, J., *Subtle Energy* (Saffron Walden: C. W. Daniel, 1987).
Davidson, J., *The Web of Life: Life Force; The Energetic Constitution of Man and the Neuro-Endocrine Connection* (Saffron Walden: C. W. Daniel, 1988).

Davidson, J., *The Secret of the Creative Vacuum: Man and the Energy Dance* (Saffron Walden: C.W. Daniel, 1989).

Dawkins, R., *The Selfish Gene* (Oxford: Oxford University Press, 1989).

Delanoy, D. and Sah, S., 'Cognitive and psychological psi responses in remote positive and neutral emotional states', in R. Bierman (ed.) *Proceedings of Presented Papers*, American Parapsychological Association, 37th Annual Convention, University of Amsterdam, 1994.

Del Giudice, E., 'The roots of cosmic wholeness are in quantum theory', *Frontier Science: An Electronic Journal*, 1997; 1(1).

Del Giudice, E. and Preparata, G., 'Water as a free electric dipole laser', *Physical Review Letters*, 1988; 61:1085-88.

Del Giudice, E. *et al.*, 'Electromagnetic field and spontaneous symmetry breaking in biological matter', *Nuclear Physics*, 1983; B275(F517): 185-99.

deLange deKlerk, E. S. M. and Bloomer, J., 'Effect of homoeopathic medicine on daily burdens of symptoms in children with recurrent upper respiratory tract infections', *British Medical Journal*, 1994; 309:1329-32.

Demangeat, L. *et al.*, 'Modifications des temps de relaxation RMN à 4MHz des protons du solvant dans les très hautes dilutions salines de silice/lactose', *Journal of Medical Nuclear Biophysics*, 1992; 16:135-45.

Dennett, D. C., *Consciousness Explained* (London: Allen Lane/Penguin, 1991).

DeValois, R. and DeValois, K., 'Spatial vision', *Annual Review of Psychology*, 1980: 309-41.

DeValois, R. and DeValois, K., *Spatial Vision* (Oxford: Oxford University Press, 1988).

DiChristina, M., 'Star travelers', *PopularScience*, 1999, June: 54-9.

Dillbeck, M. C. *et al.*, 'The Transcendental Meditation program and crime rate change in a sample of 48 cities', *Journal of Crime and Justice*, 1981; 4: 25-45.

Dobyns, Y. H., 'Combination of results from multiple experiments', Princeton Engineering Anomalies Research, *PEAR Technical Note* 97008, October 1997.

Dobyns, Y. H. *et al.*, 'Response to Hansen, Utts and Markwick: statistical and methodological problems of the PEAR remote viewing (sic) experiments', *Journal of Parapsychology*, 1992; 56:115-146.

Dossey, L., *Space, Time and Medicine* (Boston, Mass.: Shambhala, 1982).

Dossey, L., *Recovering the Soul: A Scientific and Spiritual Search* (New York: Bantam, 1989).

Dossey, L., *Healing Words: The Power of Prayer and the Practice of Medicine* (San Francisco: HarperSanFrancisco, 1993).

Dossey, L., *Prayer Is Good Medicine: How to Reap the Healing Benefits of Prayer* (San Francisco: HarperSan Francisco, 1996).

Dossey, L., *Be Careful What You Pray For . . . You Just Might Get It: What We Can Do About the Unintentional Effect of Our Thoughts, Prayers, and Wishes* (San Francisco: HarperSan Francisco, 1998).

Dossey, L., *Reinventing Medicine: Beyond Mind–Body to a New Era of Healing* (San Francisco: HarperSan Francisco, 1999).

DuBois, D. M. (ed.), *CASYS '99*: Third International Conference on Computing Anticipatory Systems (Liège, Belgium: CHAOS, 1999).

DuBois, D. M. (ed.), *CASYS 2000*: Fourth International Conference on Computing Anticipatory Systems (Liège, Belgium: CHAOS, 2000).

Dumitrescu, I. F., *Electrographic Imaging in Medicine and Biology: Electrographic Methods in Medicine and Biology* J. Kenyon (ed.), C. A. Galia (trans.) (Sudbury, Suffolk: Neville Spearman, 1983).

Dunne, B. J., 'Co-operator experiments with an REG device', Princeton Engineering Anomalies Research, *PEAR Technical Note* 91005, December 1991.

Dunne, B. J., 'Gender differences in human/machine anomalies', *Journal of Scientific Exploration*, 1998; 12(1): 3-55.

Dunne, B. and Bisaha, J., 'Precognitive remove viewing in the Chicago area: a replication of the Stanford experiment', *Journal of Parapsychology*, 1979; 43:17-30.

Dunne, B. J. and Jahn, R. G., 'Experiments in remote human/machine interaction', *Journal of Scientific Exploration*, 1992; 6(4): 311-32.

Dunne, B. J. and Jahn, R. G., 'Consciousness and anomalous physical phenomena, Princeton Engineering Anomalies Research, School of Engineering/Applied Science, *PEAR Technical Note* 95004, May 1995.

Dunne, B. J. *et al.*, 'Precognitive remote perception', Princeton Engineering Anomalies Research, *PEAR Technical Note* 83003, August 1983.

Dunne, B. J. *et al.*, 'Operator-related anomalies in a random mechanical cascade', *Journal of Scientific Exploration*, 1988; 2(2): 155-79.

Dunne, B. J. *et al.*, 'Precognitive remote perception III: complete binary data base with analytical refinements', Princeton Engineering Anomalies Research, *PEAR Technical Note* 89002, August 1989.

Dunne, J. W., *An Experiment in Time* (London: Faber, 1926).

Dziemidko, H. E., *The Complete Book of Energy Medicine* (London: Gaia, 1999).

Endler, P. C. *et al.*, 'The effect of highly diluted agitated thyroxine on the climbing activity of frogs', *Veterinary and Human Toxicology*, 1994; 36: 56-9.

Endler, P. C. *et al.*, 'Transmission of hormone information by non-molecular means', *FASEB Journal*, 1994; 8: A400(abs).

Ernst, E. and White, A., *Acupuncture: A Scientific Appraisal* (Oxford: Butterworth-Heinemann, 1999).

Ertel, S., 'Testing ESP leisurely: report on a new methodological paradigm', paper presented at the 23rd International SPR Conference, Durham, UK, September 3-5, 1999.

Feynman, R. P., *Six Easy Pieces: The Fundamentals of Physics Explained* (Harmondsworth: Penguin, 1998).

Forward, R., 'Extracting electrical energy from the vacuum by cohesion of charged foliated conductors', *Physical Review B*, 1984; 30:1700.

Fox, M. and Sheldrake, R., *The Physics of Angels: Exploring the Realm Where Science and Spirit Meet* (San Francisco: HarperSanFrancisco, 1996).

Frayn, M., *Copenhagen* (London: Methuen, 1998).

Frey, A. H., 'Electromagnetic field interactions with biological systems', *FASEB Journal*, 1993; 7: 272.

Fröhlich, H., 'Long-range coherence and energy storage in biological systems', *International Journal of Quantum Chemistry*, 1968; 2: 641-49.

Fröhlich, H., 'Evidence for Bose condensation-like excitation of coherent modes in biological systems', *Physics Letters*, 1975; 51A: 21.

Galland, L., *The Four Pillars of Healing* (New York: Random House, 1997).

Gariaev, P. P. *et al.*, 'The DNA-wave biocomputer', paper presented at CASYS 2000: Fourth International Conference on Computing Anticipatory Systems, Liège, Belgium, August 9-14, 2000.

Gerber, R., *Vibrational Medicine: New Choices for Healing Ourselves* (Santa Fe: Bear, 1988).

Gleick, J., *Chaos: Making a New Science* (London: Cardinal, 1987).

Grad, B., 'Some biological effects of "laying-on of hands": a review of experiments with animals and plants', *Journal of the American Society for Psychical Research*, 1965; 59:95-127.

Grad, B., 'Healing by the laying on of hands; review of experiments and implications', *Pastoral Psychology*, 1970; 21:19-26.

Grad, B., 'Dimensions in "Some biological effects of the laying on of hands" and their implications', in H. A. Otto and J. W. Knight (eds), *Dimensions in Wholistic Healing: New Frontiers in the Treatment of the Whole Person* (Chicago: Nelson-Hall, 1979): 199-212.

Grad, B. *et al.*, 'The influence of an unorthodox method of treatment on wound healing in mice', *International Journal of Parapsychology* 1963; 3(5): 24.

Graham, H., *Soul Medicine: Restoring the Spirit to Healing* (London: Newleaf, 2001).

Green, B., *The Elegant Universe: Superstrings, Hidden Dimensions and the Quest for the Ultimate Theory* (London: Vintage, 2000).

Green, E. E., 'Copper wall research psychology and psychophysics: subtle energies and energy medicine: emerging theory and practice', *Proceedings*, First Annual Conference, International Society for the Study of Subtle Energies and Energy Medicine (ISSSEEM), Boulder, Colo., June 21-25, 1991.

Greenfield, S. A., *Journey to the Centers of the Mind: Toward a Science of Consciousness* (New York: W. H. Freeman, 1995).

Greyson, B., 'Distance healing of patients with major depression', *Journal of Scientific Exploration*, 1996; 10(4): 447-65.

Goodwin, B., *How the Leopard Changed Its Spots: The Evolution of Complexity* (London: Phoenix, 1994).

Grinberg-Zylberbaum, J. and Ramos, J., 'Patterns of interhemisphere correlations during human communication', *International Journal of Neuroscience*, 1987; 36: 41-53.

Grinberg-Zylberbaum, J. *et al.*, 'Human communication and the electrophysiological activity of the brain', *Subtle Energies*, 1992; 3(3): 25-43.

Gribbin, J., *Almost Everyone's Guide to Science* (London: Phoenix, 1999).

Gribbin, J., *Q Is for Quantum: Particle Physics from A to Z* (London: Phoenix Giant, 1999).

Hagelin, J. S. *et al.*, 'Effects of group practice of the Transcendental Meditation Program on preventing violent crime in Washington DC: results of the National Demonstration Project, June-July, 1993', *Social Indicators Research*, 1994; 47:153-201.

Haisch, B., 'Brilliant disguise: light, matter and the Zero Point Field', *Science and Spirit*, 1999; 10: 30-1.

Haisch, B. M. and Rueda, A., 'A quantum broom sweeps clean', *Mercury: The Journal of the Astronomical Society of the Pacific*, 1996; 25(2): 12-15.

Haisch, B. M. and Rueda, A., 'The Zero Point Field and inertia', presented at Causality and Locality in Modern Physics and Astronomy: Open Questions and Possible Solutions, A symposium to honor Jean-Pierre Vigier, York University, Toronto, August 25-29, 1997.

Haisch, B. M. and Rueda, A., 'The Zero Point Field and the NASA challenge to create the space drive', presented at Breakthrough Propulsion Physics workshop, NASA Lewis Research Center, Cleveland, Ohio, August 12-14, 1997.

Haisch, B. M. and Rueda, A., 'An electromagnetic basis for inertia and gravitation: what are the implications for twenty-first century physics and technology?', presented at

Space Technology and Applications International Forum – 1998, cosponsored by NASA, DOE & USAF, Albuquerque, NM, January 25-29, 1998.

Haisch, B. M. and Rueda, A., 'Progress in establishing a connection between the electromagnetic zero point field and inertia', presented at Space Technology and Applications International Forum – 1999, cosponsored by NASA, DOE & USAF, Albuquerque, NM, January 31 to February 4, 1999.

Haisch, B. M. and Rueda, A., 'On the relation between zero-point-field induced inertial mass and the Einstein-deBroglie formula', *Physics Letters A* (in press during research).

Haisch, B., Rueda, A. and Puthoff, H. E., 'Beyond $E=mc^2$: a first glimpse of a universe without mass', *Sciences*, November/December 1994: 26-31.

Haisch, B., Rueda, A. and Puthoff, H. E., 'Inertia as a zero-point-field Lorentz force', *Physical Review A*, 1994; 49(2): 678-94.

Haisch, B., Rueda, A. and Puthoff, H. E., 'Physics of the zero point field: implications for inertia, gravitation and mass', *Speculations in Science and Technology*, 1997; 20: 99-114.

Haisch, B., Rueda, A. and Puthoff, H. E., 'Advances in the proposed electromagnetic zero-point-field theory of inertia', paper presented at AIAA 98-3143, Advances ASME/SAE/ASEE Joint Propulsion Conference and Exhibit, Cleveland, Ohio, July 13-15, 1998.

Hall, N., *The New Scientist Guide to Chaos* (Harmondsworth: Penguin, 1992).

Hameroff, S. R., *Ultimate Computing: Biomolecular Consciousness and Nanotechnology* (Amsterdam: North Holland, 1987).

Haraldsson, E. and Thorsteinsson, T., 'Psychokinetic effects on yeast: an exploratory experiment', in W. G. Roll, R. L. Morris and J. D. Morris (eds), *Research in Parapsychology* (Metuchen, NJ: Scarecrow Press, 1972): 20-21.

Harrington, A. (ed.), *The Placebo Effect: An Interdisciplinary Exploration* (Cambridge, Mass.: Harvard University Press, 1997).

Harris. W. S. *et al.*, 'A randomized, controlled trial of the effects of remote, intercessory prayer on outcomes in patients admitted to the coronary care unit', *Archives of Internal Medicine*, 1999; 159(19): 2273-78.

Hawking, S., *A Brief History of Time: From the Big Bang to Black Holes* (London: Bantam Press, 1988).

Hill, A., 'Phantom limb pain: a review of the literature on attributes and potential mechanisms', www.stir.ac.uk.

Ho, Mae-Wan, 'Bioenergetics and the coherence of organisms', *Neuronetwork World*, 1995; 5: 733-50.

Ho, Mae-Wan, 'Bioenergetics and Biocommunication', in R. Cuthbertson *et al.* (eds), *Computation in Cellular and Molecular Biological Systems* (Singapore: World Scientific, 1996): 251-64.

Ho, Mae-Wan, *The Rainbow and the Worm: The Physics of Organisms* (Singapore: World Scientific, 1999).

Hopcke, R. H., *There Are No Accidents: Synchronicity and the Stories of Our Lives* (New York: Riverhead, 1997).

Horgan, J., *The End of Science: Facing the Limits of Knowledge in the Twilight of the Scientific Age* (London: Abacus, 1998).

Hunt, V. V., *Infinite Mind: The Science of Human Vibrations* (Malibu, Calif.: Malibu, 1995).

Hyvarien, J. and Karlssohn, M., 'Low-resistance skin points that may coincide with acupuncture loci', *Medical Biology*, 1977; 55: 88-94, as quoted in the *New England*

Journal of Medicine, 1995; 333(4): 263.

Ibison, M., 'Evidence that anomalous statistical influence depends on the details of random process', *Journal of Scientific Exploration*, 1998; 12(3): 407-23.

Ibison, M. and Jeffers, S., 'A double-slit diffraction experiment to investigate claims of consciousness-related anomalies', *Journal of Scientific Exploration*, 1998; 12(4): 543-50.

Insinna, E., 'Synchronicity and coherent excitations in microtubules', *Nanobiology*, 1992; 1:191-208.

Insinna, E., 'Ciliated cell electrodynamics: from cilia and flagella to ciliated sensory systems', in A. Malhotra (ed.) *Advances in Structural Biology* (Stamford, Connecticut: JAI Press, 1999): 5.

Jacobs, J., 'Homoeopathic treatment of acute childhood diarrhoea', *British Homoeopathic Journal*, 1993; 82: 83-6.

Jahn, R. G., 'The persistent paradox of psychic phenomena: an engineering perspective', *IEEE Proceedings of the IEEE*, 1982; 70(2): 136-70.

Jahn, R., 'Physical aspects of psychic phenomena', *Physics Bulletin*, 1988; 39: 235-37.

Jahn, R. G., 'Acoustical resonances of assorted ancient structures', *Journal of the Acoustical Society of America*, 1996; 99(2): 649-58.

Jahn, R. G., 'Information, consciousness, and health', *Alternative Therapies*, 1996; 2(3): 32-8.

Jahn, R., 'A modular model of mind/matter manifestations', *PEAR Technical Note* 2001.01, May 2001 (abstract).

Jahn, R. G. and Dunne, B. J., 'On the quantum mechanics of consciousness with application to anomalous phenomena', *Foundations of Physics*, 1986; 16(8): 721-72.

Jahn, R. G. and Dunne, B. J., *Margins of Reality: The Role of Consciousness in the Physical World* (London: Harcourt Brace Jovanovich, 1987).

Jahn, R. and Dunne, B., 'Science of the subjective', *Journal of Scientific Exploration*, 1997; 11(2): 201-24.

Jahn, R. G. and Dunne, B. J., 'ArtREG: a random event experiment utilizing picture-preference feedback', *Journal of Scientific Exploration*, 2000; 14(3): 383-409.

Jahn, R. G. *et al.*, 'Correlations of random binary sequences with prestated operator intention: a review of a 12-year program', *Journal of Scientific Exploration*, 1997; 11: 345-67.

Jaynes, J., *The Origin of Consciousness in the Breakdown of the Bicameral Mind* (Harmondsworth: Penguin, 1990).

Jibu, M. and Yasue, K., 'A physical picture of Umezawa's quantum brain dynamics', in R. Trappl (ed.) *Cybernetics and Systems Research, '92* (Singapore: World Scientific, 1992).

Jibu, M. and Yasue, K., 'The basis of quantum brain dynamics', in K. H. Pribram (ed.) *Proceedings of the First Appalachian Conference on Behavioral Neurodynamics*, Radford University, September 17-20, 1992 (Radford: Center for Brain Research and Informational Sciences, 1992).

Jibu, M. and Yasue, K., 'Intracellular quantum signal transfer in Umezawa's quantum brain dynamics', *Cybernetic Systems International*, 1993; 1(24): 1-7.

Jibu, M. and Yasue, K., 'Introduction to quantum brain dynamics', in E. Carvallo (ed.), *Nature, Cognition and System III* (London: Kluwer Academic, 1993).

Jibu, M. and Yasue, K., 'The basis of quantum brain dynamics', in K. H. Pribram (ed.), *Rethinking Neural Networks: Quantum Fields and Biological Data* (Hillsdale, NJ: Lawrence Erlbaum, 1993): 121-45.

Jibu, M. *et al.*, 'Quantum optical coherence in cytoskeletal microtubules: implications for brain function', *BioSystems*, 1994; 32: 95-209.

Jibu, M. *et al.*, 'From conscious experience to memory storage and retrieval: the role of quantum brain dynamics and boson condensation of evanescent photons', *International Journal of Modern Physics B*, 1996; 10(13/14): 1735-54.

Kaplan, G. A. *et al.*, 'Social connections and morality from all causes and from cardiovascular disease: perspective evidence from eastern Finland, *American Journal of Epidemiology*, 1988; 128: 370-80.

Katchmer, G. A. Jr, *The Tao of Bioenergetics* (Jamaica Plain, Mass.: Yang's Martial Arts Association, 1993).

Katra, J. and Targ, R., *The Heart of the Mind: How to Experience God Without Belief* (Novato, Calif.: New World Library, 1999).

Kelly, M. O. (ed.), *The Fireside Treasury of Light: An Anthology of the Best in New Age Literature* (London: Fireside/Simon & Schuster, 1990).

Kiesling, S., 'The most powerful healing God and women can come up with', *Spirituality and Health*, 1999; winter: 22-7.

King, J. *et al.*, 'Spectral density maps of receptive fields in the rat's somatosensory cortex', in *Origins: Brain and Self Organization* (Hillsdale, NJ: Lawrence Erlbaum, 1995).

Klebanoff, N. A. and Keyser, P. K., 'Menstrual synchronization: a qualitative study', *Journal of Holistic Nursing*, 1996; 14(2): 98-114.

Krishnamurti and Bohm, D., *The Ending of Time: Thirteen Dialogues* (London: Victor Gollancz, 1991).

Lafaille, R. and Fulder, S. (eds), *Towards a New Science of Health* (London: Routledge, 1993).

Laszlo, E., *The Interconnected Universe: Conceptual Foundations of Transdisciplinary Unified Theory* (Singapore: World Scientific, 1995).

Laughlin, C. D., 'Archetypes, neurognosis and the quantum sea', *Journal of Scientific Exploration*, 1996; 10: 375-400.

Lechleiter, J. *et al.*, 'Spiral waves: spiral calcium wave propagation and annihilation in Xenopus laevis oocytes', *Science*, 1994; 263: 613.

Lee, R. H., *Bioelectric Vitality: Exploring the Science of Human Energy* (San Clemente, Calif.: China Healthways Institute, 1997).

Lessell, C. B., *The Infinitesimal Dose: The Scientific Roots of Homeopathy* (Saffron Walden: C. W. Daniel, 1994).

Levitt, B. B., *Electromagnetic Fields; A Consumer's Guide to the Issues and How to Protect Ourselves* (New York: Harcourt Brace, 1995).

Liberman, J., *Light: Medicine of the Future* (Santa Fe, NM: Bear, 1991).

Light, M., *Full Moon* (London: Jonathan Cape, 1999).

Liquorman, W. (ed.), *Consciousness Speaks: Conversations with Ramesh S. Balsekar* (Redondo Beach, Calif.: Advaita Press, 1992).

Lorimer, D. (ed.), *The Spirit of Science: From Experiment to Experiment* (Edinburgh: Floris, 1998).

Lovelock, J., *Gaia: A New look at Life on Earth* (Oxford: Oxford University Press, 1979).

Loye, D., *An Arrow Through Chaos* (Rochester, Vt.: Park Street Press, 2000).

Loye, D., *Darwin's Lost Theory of Love: A Healing Vision for the New Century* (Lincoln, Neb.: iUniverse.com, Inc., 2000).

Marcer, P. J., 'A quantum mechanical model of evolution and consciousness', *Proceedings of the 14th International Congress of Cybernetics*, Namur, Belgium, August 22-26, 1995, Symposium XI: 429-34.

Marcer, P. J., 'Getting quantum theory off the rocks', *Proceedings of the 14th International Congress of Cybernetics*, Namur, Belgium, August, 22-26, 1995, Symposium XI: 435-40.

Marcer, P. J., 'The jigsaw, the elephant and the lighthouse', *ANPA 20 Proceedings*, 1998, 93-102.

Marcer, P. J. and Schempp, W., 'Model of the neuron working by quantum holography', *Informatica*, 1997; 21: 519-34.

Marcer, P. J. and Schempp, W., 'The model of the prokaryote cell as an anticipatory system working by quantum holography', *Proceedings of the First International Conference on Computing Anticipatory Systems*, Liège, Belgium, August 11-15, 1997.

Marcer, P. J. and Schempp, W., 'The model of the prokaryote cell as an anticipatory system working by quantum holography', *International Journal of Computing Anticipatory Systems*, 1997; 2: 307-15.

Marcer, P. J. and Schempp, W., 'The brain as a conscious system', *International Journal of General Systems*, 1998; 27(1-3): 231-48.

Mason, K., *Medicine for the Twenty-First Century: The Key to Healing with Vibrational Medicine* (Shaftesbury, Dorset: Element, 1992).

Master, F. J., 'A study of homeopathic drugs in essential hypertension', *British Homoeopathic Journal*, 1987; 76: 120-1.

Matthews, D. A., *The Faith Factor: Proof of the Healing Power of Prayer* (New York: Viking, 1998).

Matthews, R., 'Does empty space put up the resistance?', *Science*, 1994; 263: 613.

Matthews, R., 'Nothing like a vacuum', *New Scientist*, February 25, 1995: 30-33.

Matthews, R., 'Vacuum power could clean up', *Sunday Telegraph*, December 31, 1995.

McKie, R., 'Scientists switch to warp drive as sci-fi energy source is tapped', *Observer*, January 7, 2001.

McMoneagle, J., *Mind Trek: Exploring Consciousness, Time, and Space through Remote Viewing* (Charlottesville, Va.: Hampton Road, 1997).

McMoneagle, J., *The Ultimate Time Machine: A Remote Viewer's Perception of Time, and Predictions for the New Millennium* (Charlottesville, Va.: Hampton Road, 1998).

Miller, R. N., 'Study on the effectiveness of remote mental healing', *Medical Hypotheses*, 1982; 8: 481-90.

Milonni, P.W., 'Semi-classical and quantum electrodynamical approaches in nonrelativistic radiation theory', *Physics Reports*, 1976; 25:1-8.

Mims, C., *When We Die* (London: Robinson, 1998).

Mitchell, E., *The Way of the Explorer: An Apollo Astronaut's Journey Through the Material and Mystical Worlds* (London: G. P. Putnam, 1996).

Mitchell, E., 'Nature's mind', keynote address to CASYS 1999: Third International Conference on Computing Anticipatory Systems, August 8, 1999 (Liège, Belgium: CHAOS, 1999).

Moody, R. A. Jr, *The Light Beyond* (New York: Bantam, 1989).

Morris, R. L. *et al.*, 'Comparison of the sender/no sender condition in the ganzfeld', in N. L. Zingrone (ed.), *Proceedings of Presented Papers*, 38th Annual Parapsychological Association Convention (Fairhaven, Mass.: Parapsychological Association).

Moyers, W., *Healing and the Mind* (London: Aquarian/Thorsons, 1993).

Murphy, M., *The Future of the Body: Explorations into the Further Evolution of Human Nature* (Los Angeles: Jeremy P. Tarcher, 1992).

Nash, C. B., 'Psychokinetic control of bacterial growth?', *Journal of the American Society for Psychical Research*, 1982; 51: 217-21.

Nelson, R. D., 'Effect size per hour: a natural unit for interpreting anomalous experiments', Princeton Engineering Anomalies Research, School of Engineering/Applied Science, *PEAR Technical Note* 94003, September 1994.

Nelson, R., 'FieldREG measurements in Egypt: resonant consciousness at sacred sites', Princeton Engineering Anomalies Research, School of Engineering/Applied Science, *PEAR Technical Note* 97002, July 1997.

Nelson, R., 'Wishing for good weather: a natural experiment in group consciousness', *Journal of Scientific Exploration*, 1997; 11(1): 47-58.

Nelson, R. D., 'The physical basis of intentional healing systems', Princeton Engineering Anomalies Research, School of Engineering/Applied Science, *PEAR Technical Note* 99001, January 1999.

Nelson, R. D. and Radin, D. I., 'When immovable objections meet irresistible evidence', *Behavioral and Brain Sciences*, 1987; 10: 600-601.

Nelson, R. D. and Radin, D. I., 'Statistically robust anomalous effects: replication in random event generator experiments', in L. Henckle and R. E. Berger (eds) *RIP 1988* (Metuchen, NJ: Scarecrow Press, 1989).

Nelson, R. D. and Mayer, E. L., 'A FieldREG application at the San Francisco Bay Revels, 1996', as reported in D. Radin, *The Conscious Universe: The Scientific Truth of Psychic Phenomena* (New York: HarperEdge, 1997): 171.

Nelson, R. D. *et al.*, 'A linear pendulum experiment: effects of operator intention on damping rate', *Journal of Scientific Exploration*, 1994; 8(4): 471-89.

Nelson, R. D. *et al.*, 'FieldREG anomalies in group situations', *Journal of Scientific Exploration*, 1996; 10(1): 111-41.

Nelson, R. D. *et al.*, 'FieldREGII: consciousness field effects: replications and explorations', *Journal of Scientific Exploration*, 1998; 12(3): 425-54.

Nelson, R. *et al.*, 'Global resonance of consciousness: Princess Diana and Mother Teresa', *Electronic Journal of Parapsychology*, 1998.

Ness, R. M. and Williams, G. C., *Evolution and Healing: The New Science of Darwinian Medicine* (London: Phoenix, 1996).

Nobili, R., 'Schrödinger wave holography in brain cortex', *Physical Review* A, 1985; 32: 3618-26.

Nobili, R., 'Ionic waves in animal tissues', *Physical Review* A, 1987; 35:1901-22.

Nuland, S. B., *How We Live: The Wisdom of the Body* (London: Vintage, 1997).

Odier, M., 'Psycho-physics: new developments and new links with science', paper presented at the Fourth Biennial European Meeting of the Society for Scientific Exploration, Valencia, October 9-11, 1998.

Ornstein, R. and Swencionis, C. (eds), *The Healing Brain: A Scientific Reader* (New York: Guilford Press, 1990).

Orme-Johnson, W. *et al.*, 'International peace project in the Middle East: the effects of the Maharishi technology of the unified field', *Journal of Conflict Resolution*, 1988; 32: 776-812.

Ostrander, S. and Schroeder, L., *Psychic Discoveries* (New York: Marlowe, 1997).

Pascucci, M. A. and Loving, G. L., 'Ingredients of an old and healthy life: centenarian perspective', *Journal of Holistic Nursing*, 1997; 15:199-213.

Penrose, R., *The Emperor's New Mind: Concerning Computers, Minds and The Laws of Physics* (Oxford: Oxford University Press, 1989).

Penrose, R., *Shadows of the Mind: A Search for the Missing Science of Consciousness* (London: Vintage, 1994).

Peoc'h, R., 'Psychokinetic action of young chicks on the path of an illuminated source', *Journal of Scientific Exploration*, 1995; 9(2): 223.

Pert, C., *Molecules of Emotion: Why You Feel the Way You Feel* (London: Simon & Schuster, 1998).

Pinker, S., *How the Mind Works* (Harmondsworth: Penguin, 1998).

Pomeranz, B. and Stu, G., *Scientific Basis of Acupuncture* (New York: Springer-Verlag, 1989).

Popp, F. A., 'Biophotonics: a powerful tool for investigating and understanding life', in H. P. Dürr, F. A. Popp and W. Schommers (eds), *What is Life?* (Singapore: World Scientific), in press.

Popp, F. A. and Chang, Jiin-Ju, 'Mechanism of interaction between electromagnetic fields and living systems.' *Science in China (Series C)*, 2000; 43: 507-18.

Popp, F. A., Gu, Qiao and Li, Ke-Hsueh, 'Biophoton emission: experimental background and theoretical approaches', *Modern Physics Letters B*, 1994; 8(21/22): 1269-96.

Powell, A. E., *The Etheric Double and Allied Phenomena* (London: Theosophical Publishing House, 1979).

Pribram, K. H., *Languages of the Brain: Experimental Paradoxes and Principles in Neuropsychology* (New York: Brandon House, 1971).

Pribram, K. H., *Brain and Perception: Holonomy and Structure in Figural Processing* (Hillsdale, NJ: Lawrence Erlbaum, 1991).

Pribram, K. H. (ed.), *Rethinking Neural Networks: Quantum Fields and Biological Data*, Proceedings of the First Appalachian Conference on Behavioral Neurodynamics (Hillsdale, NJ: Lawrence Erlbaum, 1993).

Pribram, K. H., 'Autobiography in anecdote: the founding of experimental neuropsychology', in R. Bilder (ed.), *The History of Neuroscience in Autobiography* (San Diego, Calif.: Academic Press, 1998): 306-49.

Puthoff, H., 'Toward a quantum theory of life process', unpublished, 1972.

Puthoff, H. E., 'Experimental psi research: implication for physics', in R. Jahn (ed.), *The Role of Consciousness in the Physical World*, AAA Selected Symposia Series (Boulder, Colo.: Westview Press, 1981).

Puthoff, H. E., 'ARV (associational remote viewing) applications', in R. A. White and J. Solfvin (eds), *Research in Parapsychology 1984*, Abstracts and Papers from the 27th Annual Convention of the Parapsychological Association, 1984 (Metuchen, NJ: Scarecrow Press, 1985).

Puthoff, H., 'Ground state of hydrogen as a zero-point-fluctuation-determined state', *Physical Review D*; 1987, 35: 3266.

Puthoff, H. E., 'Gravity as a zero-point-fluctuation force', *Physical Review A*, 1989; 39(5): 2333-42.

Puthoff, H. E., 'Source of vacuum electromagnetic zero-point energy', *Physical Review A*, 1989; 40: 4857-62.

Puthoff, H., 'Where does the zero-point energy come from?', *New Scientist*, December 2, 1989: 36.

Puthoff, H., 'Everything for nothing', *New Scientist*, July 28, 1990: 52-5.

Puthoff, H. E., 'The energetic vacuum: implications for energy research', *Speculations in Science and Technology*, 1990; 13(4): 247.

Puthoff, H. E., 'Reply to comment', *Physical Review A*, 1991; 44: 3385-86.

Puthoff, H. E., 'Comment', *Physical Review A*, 1993; 47(4): 3454-55.

Puthoff, H. E., 'CIA-initiated remote viewing program at Stanford Research Institute', *Journal of Scientific Exploration*, 1996; 10(1): 63-76.

Puthoff, H., 'SETI, the velocity-of-light limitation, and the Alcubierre warp drive: an integrating overview', *Physics Essays*, 1996; 9(1): 156-8.

Puthoff, H., 'Space propulsion: can empty space itself provide a solution?', *Ad Astra*, 1997; 9(1): 42-6.

Puthoff, H. E., 'Can the vacuum be engineered for spaceflight applications? Overview of theory and experiments', *Journal of Scientific Exploration*, 1998; 12(10): 295-302.

Puthoff, H., 'On the relationship of quantum energy research to the role of metaphysical processes in the physical world', 1999, posted on www. meta-list.org.

Puthoff, H. E., 'Polarizable-vacuum (PV) representation of general relativity', September 1999, posted on Los Alamos archival website www.lanl.gov/worldview/.

Puthoff, H., 'Warp drive win? Advanced propulsion', *Jane's Defence Weekly*, July 26, 2000: 42-6.

Puthoff, H. and Targ, R., 'Physics, entropy, and psychokinesis', in L. Oteri (ed.), *Quantum Physics and Parapsychology*, Proceedings of an International Conference held in Geneva, Switzerland, August 26-27, 1974.

Puthoff, H. and Targ, R., 'A perceptual channel for information transfer over kilometer distances: historical perspective and recent research', *Proceedings of the IEEE*, 1976; 64(3): 329-54.

Puthoff, H. and Targ, R., 'Final report, covering the period January 1974-February 1975', December 1, 1975, *Perceptual Augmentation Techniques*, Part I and II, SRI Projects 3183, classified documents until July 1995.

Puthoff, H. E. *et al.*, 'Calculator-assisted PSI amplication II: use of the sequential-sampling technique as a variable-length majority vote code', in D. H. Weiner and D. I. Radin (eds), *Research in Parapsychology 1985*, Abstracts and Papers from the 28th Annual Convention of the Parapsychological Association, 1985 (Metuchen, NJ: Scarecrow Press, 1986).

Radin, D. I., *The Conscious Universe: The Scientific Truth of Psychic Phenomena* (New York: HarperEdge, 1997).

Radin, D. and Ferrari, D. C., 'Effect of consciousness on the fall of dice: a meta-analysis', *Journal of Scientific Exploration*, 1991; 5: 61-84.

Radin, D. I. and May, E. C., 'Testing the intuitive data sorting model with pseudorandom number generators: a proposed method', in D. H. Weiner and R. G. Nelson (eds), *Research in Parapsychology 1986* (Metuchen, NJ: Scarecrow Press, 1987): 109-11.

Radin, D. and Nelson, R., 'Evidence for consciousness-related anomalies in random physical systems', *Foundations of Physics*, 1989; 19(12): 1499-514.

Radin, D. and Nelson, R., 'Meta-analysis of mind-matter interaction experiments, 1959-2000', www.boundaryinstitute.org.

Radin, D. I., Rebman, J. M. and Cross, M. P., 'Anomalous organization of random events by group consciousness: two exploratory experiments', *Journal of Scientific Exploration*, 1996: 143-68.

Randles, J., *Paranormal Source Book: The Comprehensive Guide to Strange Phenomena Worldwide* (London: Judy Piatkus, 1999).

Reanney, D., *After Death: A New Future for Human Consciousness* (New York: William Morrow, 1991).

Reed, D. *et al.*, 'Social networks and coronary heart disease among Japanese men in Hawaii', *American Journal of Epidemiology*, 1983; 117: 384-96.

Reilly, D., 'Is evidence for homeopathy reproducible?', *Lancet*, 1994; 344: 1601-06.

Robinson, C. A. Jr, 'Soviets push for beam weapon', *Aviation Week*, May 2, 1977.

Rosenthal, R., 'Combining results of independent studies', *Psychological Bulletin*, 1978; 85: 185-93.

Rubik, B., *Life at the Edge of Science* (Oakland, Calif.: Institute for Frontier Science, 1996).

Rueda, A. and Haisch, B., 'Contribution to inertial mass by reaction of the vacuum to accelerated motion', *Foundations of Physics*, 1998; 28(7): 1057-107.

Ruéda, A., Haisch, B. and Cole, D. C., 'Vacuum zero-point-field pressure instability in astrophysical plasmas and the formation of cosmic voids', *Astrophysical Journal*, 1995; 445: 7-16.

Sagan, Carl, *Contact* (London: Orbit, 1997).

Sanders, P. A. Jr, *Scientific Vortex Information: An M.I.T.-Trained Scientist's Program* (Sedona, Ariz.: Free Soul, 1992).

Sardello, R., 'Facing the world with soul: disease and the reimagination of modern life', *Aromatherapy Quarterly*, 1992; 35: 13-7.

Schiff, M., *The Memory of Water: Homeopathy and the Battle of Ideas in the New Science* (London: Thorsons, 1995).

Schiff, M., 'On consciousness, causation and evolution', *Alternative Therapies*, July 1998; 4(4): 82-90.

Schiff, M. and Braud, W., 'Distant intentionality and healing: assessing the evidence', *Alternative Therapies*, 1997; 3(6): 62-73.

Schlitz, M. J. and Honorton, C., 'Ganzfeld psi performance within an artistically gifted population', *Journal of the American Society for Psychical Research*, 1992; 86(2): 83-98.

Schlitz, M. and LaBerge, S., 'Autonomic detection of remote observation: two conceptual replications', in D. J. Bierman (ed.) *Proceedings of Presented Papers*, 37th Annual Parapsychological Association Convention, Amsterdam (Fairhaven, Mass.: Parapsychological Association, 1994): 352-60.

Schlitz, M. J. and LaBerge, S., 'Covert observation increases skin conductance in subjects unaware of when they are being observed: a replication', *Journal of Parapsychology*, 1997; 61: 185-96.

Schmidt, H., 'Quantum processes predicted?', *New Scientist,* October 16, 1969: 114-15.

Schmidt, H., 'Mental influence on random events', *New Scientist and Science Journal*, June 24, 1971; 757-8.

Schmidt, H., 'Toward a mathematical theory of psi', *Journal of the American Society for Psychical Research*, 1975; 69(4): 301-319.

Schmidt, H., 'Additional affect for PK on pre-recorded targets', *Journal of Parapsychology*, 1985; 49: 229-44.

Schnabel, J., *Remote Viewers: The Secret History of America's Psychic Spies* (New York: Dell, 1997).

Schwarz, G. *et al.*, 'Accuracy and replicability of anomalous after-death communication across highly skilled mediums', *Journal of the Society for Psychical Research*, 2001; 65: 1-25.

Scott-Mumby, K., *Virtual Medicine: A New Dimension in Energy Healing* (London: Thorsons, 1999).

Senekowitsch, F. *et al.*, 'Hormone effects by CD record/replay', *FASEB Journal*, 1995; 9: A392 (abs).

Sharma, H., 'Lessons from the placebo effect', *Alternatives Therapies in Clinical Practice*, 1997; 4(5): 179-84.

Shealy, C. N., *Sacred Healing: The Curing Power of Energy and Spirituality* (Boston, Mass.: Element, 1999).

Sheldrake, R., *A New Science of Life: The Hypothesis of Formative Causation* (London: Paladin, 1987).

Sheldrake, R., 'An experimental test of the hypothesis of formative causation', *Rivista Di Diologia-Biology Forum*, 1992; 85(3/4): 431-3.

Sheldrake, R., *The Presence of the Past: Morphic Resonance and the Habits of Nature* (London: HarperCollins, 1994).

Sheldrake, R., *The Rebirth of Nature: The Greening of Science and God* (Rochester, Vt.: Park Street Press, 1994).

Sheldrake, R., *Seven Experiments That Could Change the World: A Do-It-Yourself Guide to Revolutionary Science* (London: Fourth Estate, 1995).

Sheldrake, R., 'Experimenter effects in scientific research: how widely are they neglected?', *Journal of Scientific Exploration*, 1998; 12(1): 73-8.

Sheldrake, R., 'The sense of being stared at: experiments in schools', *Journal of the Society for Psychical Research*, 1998; 62: 311-23.

Sheldrake, R., 'Could experimenter effects occur in the physical and biological sciences?', *Skeptical Inquirer*, 1998; 22(3): 57-8.

Sheldrake, R., *Dogs that Know When Their Owners Are Coming Home and Other Unexplained Powers of Animals* (London: Hutchinson, 1999).

Sheldrake, R., 'How widely is blind assessment used in scientific research?', *Alternative Therapies*, 1999; 5(3): 88-91.

Sheldrake, R., 'The "sense of being stared at" confirmed by simple experiments', *Biology Forum*, 1999; 92: 53-76.

Sheldrake, R. and Smart, P. ,'A dog that seems to know when his owner is returning: preliminary investigations', *Journal of the Society for Psychical Research*, 1998; 62: 220-32.

Sheldrake, R. and Smart, P. ,'Psychic pets: a survey in north-west England', *Journal of the Society for Psychical Research*, 1997; 68: 353-64.

Sicher, F., Targ, E. *et al.*, 'A randomized double-blind study of the effect of distant healing in a population with advanced AIDS: report of a small scale study', *Western Journal of Medicine*, 1998; 168(6): 356-63.

Sigma, R., *Ether-Technology: A Rational Approach to Gravity Control* (Kempton, Ill.: Adventures Unlimited Press, 1996).

Silver, B. L., *The Ascent of Science* (London: Solomon Press/Oxford University Press, 1998).

Snel, F. W. J., 'PK Influence on malignant cell growth research', *Letters of the University of Utrecht*, 1980; 10: 19-27.

Snel, F. W. J. and Hol, P. R., 'Psychokinesis experiments in casein induced amyloidosis of the hamster', *Journal of Parapsychology*, 1983; 5(1): 51-76.

Snellgrove, B., *The Unseen Self: Kirlian Photography Explained* (Saffron Walden: C. W. Daniel, 1996).

Solfvin, G. F., 'Psi expectancy effects in psychic healing studies with malarial mice', *European Journal of Parapsychology*, 1982; 4(2): 160-97.

Stapp, H., 'Quantum Theory and the Role of Mind in Nature; *Foundations of Physics*, 2001; 31:1465-99.

Squires, E. J., 'Many views of one world – an interpretation of quantum theory', *European Journal of Physics*, 1987; 8: 173.

Stanford, R., '"Associative activation of the unconscious" and "visualization" as methods for influencing the PK target', *Journal of the American Society for Psychical Research*, 1969; 63: 338-51.

Stevenson, I., *Children Who Remember Previous Lives* (Charlottesville, Va.: University Press of Virginia, 1987).

Stillings, D., 'The historical context of energy field concepts', *Journal of the U.S. Psychotronics Association*, 1989; 1(2): 4-8.

Talbot, M., *The Holographic Universe* (London: HarperCollins, 1996).

Targ, E., 'Evaluating distant healing: a research review', *Alternative Therapies*; 1997; 3(6): 74-8.

Targ, E., 'Research in distant healing intentionality is feasible and deserves a place on our national research agenda', *Alternative Therapies*, 1997; 3(6): 92-6.

Targ, R. and Harary, K., *The Mind Race: Understanding and Using Psychic Abilities* (New York: Villard, 1984).

Targ, R. and Katra, J., *Miracles of Mind: Exploring Nonlocal Consciousness and Spiritual Healing* (Novato, Calif.: New World Library, 1999).

Targ, R. and Puthoff, H., *Mind-Reach: Scientists Look at Psychic Ability* (New York: Delacorte Press, 1977).

Tart, C., 'Physiological correlates of psi cognition', *International Journal of Parapsychology* 1963; 5: 375-86.

Tart, C., 'Psychedelic experiences associated with a novel hypnotic procedure: mutual hypnosis', in C. T. Tart (ed.) *Altered States of Consciousness* (New York: John Wiley, 1969): 291-308.

'"The truth about psychics" - what the scientists are saying ...', *The Week*, March 17, 2001.

Thomas, Y., 'Modulation of human neutrophil activation by "electronic" phorbol myristate acetate (PMA)', *FASEB Journal*, 1996; 10: A1479.

Thomas, Y. *et al.*, 'Direct transmission to cells of a molecular signal (phorbol myristate acetate, PMA) via an electronic device, *FASEB Journal*, 1995; 9: A227.

Thompson Smith, A., *Remote Perceptions: Out-of-Body Experiences, Remote Viewing and Other Normal Abilities* (Charlottesville, Va.: Hampton Road, 1998).

Thurnell-Read, J., *Geopathic Stress: How Earth Energies Affect Our Lives* (Shaftesbury, Dorset: Element, 1995).

Tiller, W. A., 'What are subtle energies', *Journal of Scientific Exploration*, 1993; 7(3): 293-304.

Tsong, T. Y., 'Deciphering the language of cells', *Trends in Biochemical Sciences*, 1989; 14: 89-92.

Utts, J., 'An assessment of the evidence for psychic functioning', *Journal of Scientific Exploration*, 1996; 10: 3-30.

Utts, J. and Josephson, B. D., 'The paranormal: the evidence and its implications for consciousness' (originally published in slighter shorter form), *New York Times Higher Education Supplement*, April 5, 1996: v.

Vaitl, D., 'Anomalous effects during Richard Wagner's operas', paper presented at the Fourth Biennial European Meeting of the Society for Scientific Exploration, Valencia, Spain, October 9-11, 1998.

Vincent, J. D., *The Biology of Emotions*, J. Hughes (trans) (Oxford: Basil Blackwell, 1990).

Vithoulkas, G., *A New Model for Health and Disease* (Mill Valley, Calif.: Health and Habitat, 1991).

Wallach, H., 'Consciousness studies: a reminder', paper presented at the Fourth Biennial European Meeting of the Society for Scientific Exploration, Valencia, Spain, October 9-11, 1998.

Walleczek, J., 'The frontiers and challenges of biodynamics research', in Jan Walleczek (ed.), *Self-organized Biological Dynamics and Nonlinear Control: Toward Understanding Complexity, Chaos and Emergent Function in Living Systems* (Cambridge: Cambridge University Press, 2000).

Weiskrantz, L., *Consciousness Lost and Found: A Neuropsychological Exploration* (Oxford: Oxford University Press, 1997).

Wezelman, R. *et al.*, 'An experimental test of magic: healing rituals', *Proceedings of Presented Papers*, 37th Annual Parapsychological Association Convention, San Diego, Calif. (Fairhaven, Mass.: Parapsychological Association, 1996): 1-12.

Whale, J., *The Catalyst of Power: The Assemblage Point of Man* (Forres, Scotland: Findhorn Press, 2001).

White, M., *The Science of the X-Files* (London: Legend, 1996).

'Why atoms don't collapse', *New Scientist*, July 9, 1997: 26.

Williamson, T., 'A sense of direction for dowsers?', *New Scientist*, March 19, 1987: 40-3.

Wolf, F. A., *The Body Quantum: The New Physics of Body Mind, and Health* (London: Heinemann, 1987).

Wolfe, T., *The Right Stuff* (London: Picador, 1990).

Youbicier-Simo, B. J. *et al.*, 'Effects of embryonic bursectomy and *in ovo* administration of highly diluted bursin on an adrenocorticotropic and immune response to chickens', *International Journal of Immunotherapy*, 1993; IX: 169-80.

Zeki, S., *A Vision of the Brain* (Oxford: Blackwell Scientific, 1993).

Zohar, D. *The Quantum Self* (London: Flamingo, 1991).

INDEX